The Struggle for Maize

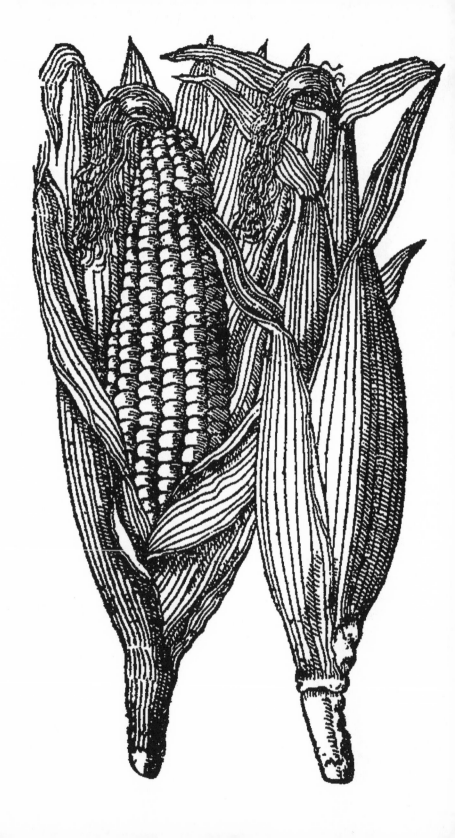

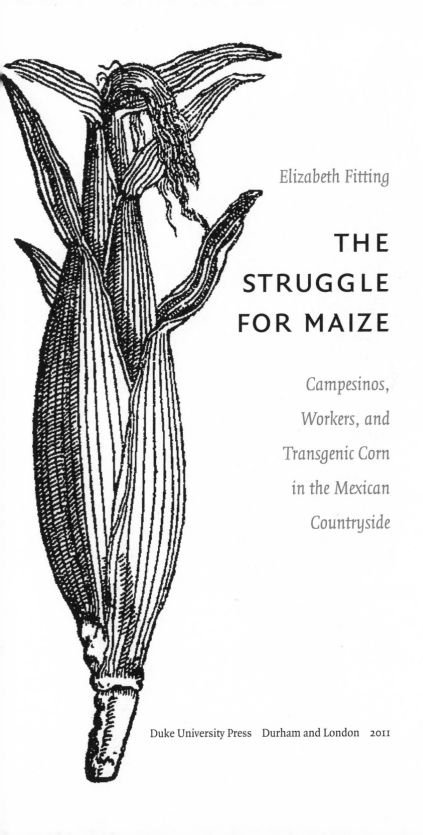

Elizabeth Fitting

THE STRUGGLE FOR MAIZE

Campesinos,
Workers, and
Transgenic Corn
in the Mexican
Countryside

Duke University Press   Durham and London   2011

© 2011 Duke University Press
All rights reserved.
Printed in the United States of
America on acid-free paper ∞
Designed by C. H. Westmoreland
Typeset in Quadraat by
Keystone Typesetting, Inc.
Library of Congress Cataloging-in-
Publication Data appear on the last
printed page of this book.

For my parents

Contents

Tables

Acknowledgments

I am indebted to many people in the Tehuacán area. First and foremost, I am grateful to the people of the southern valley who shared their insights and experiences with me. I particularly want to thank Regino Melchor Jiménez Escamilla, who helped me conduct interviews, discussed many details of this book, and offered me his friendship. My stay in the valley was enriched by his humor and intellect. I also thank my English students from the valley, some of them migrants, for sharing their stories. Zoilo Noel Guzmán Herrera spontaneously offered to become my research assistant during one of his visits home, and I thank him for his company. On my first trip to San José, Celedonio and Irma Zamora extended to me their hospitality, for which I am grateful. In Tehuacán, I thank Juan Manuel Gámez Andrade, then director of the Tehuacán Archives, and Angel Barroso Díaz, who granted me access to their private archives of Tehuacán newspapers.

Over the years I have benefited from conversations with scholars, researchers, and friends based in Mexico. A special thanks goes to Ricardo F. Macip, who suggested the valley to me as a field site and offered perceptive comments and friendship throughout the evolution of this project. I am grateful to Antonio Serratos and María Colin for so patiently explaining various aspects of transgenic corn regulation on numerous occasions; David Barkin for his knack for asking the right questions; Leigh Binford for his editorial comments on an early summary of my research; Nancy Churchill for her always engaging discussion; Edit Antal and Yolanda Massieu Trigo for sharing their considerable knowledge on GM regulation and politics; and to all the people who agreed to be interviewed about corn politics and GM regulation over the years. For their hospitality I thank the Macip family in Puebla, Eduardo in Mexico City, and Rose in Tehuacán; and for their company I thank Efraín, Marisela, María, Mahua, and especially Antonio Ortega in Cuernavaca.

While studying at the New School for Social Research I had the good fortune of working with professors who were both challenging and encouraging. I would especially like to thank Deborah Poole for her insightful comments and support. I also benefited from the guidance of Kate Crehan, Rayna Rapp, and William Roseberry; and my project was enriched by discussions with Erin Koch, Lauren Martin, Richard Wells, and Catherine Ziegler.

During the process of research and writing, various individuals and institutions provided different kinds of assistance for the project. I am grateful to David Cleveland, Cathrine Degnen, Birgit Müller, Gerardo Otero, and Gavin A. Smith for their comments at different stages of thinking through the research and writing; to Michelle Capistran Early for her assistance in the field; to Ignacio Chapela for his helpful interview; to my copy editor, Fred Kameny, and the manuscript's anonymous reviewers for their very constructive comments; to Citlalli Reyes for transcribing a couple of interviews; to Paul Lobue for his work on the maps; and to Natasha Hanson and Mafe Rueda for their help with the manuscript's layout.

I would like to extend my thanks to my colleagues at Dalhousie University, particularly to Lindsay Dubois for always being available for feedback and brainstorming; Kregg Hetherington, Christina Holmes, Matt Schnurr, and Emma Whelan for their comments on a section of the manuscript; and Liesl Gambold and Pauline Gardiner Barber for their advice and encouragement.

I am grateful to the Wenner-Gren Foundation for Anthropological Research for generously supporting portions of both the research and writing stages of this project. The New School and Dalhousie University also funded portions of the research. I would also like to acknowledge the permission of Springer for reprinting parts of "Importing Corn, Exporting Labor: The Neoliberal Corn Regime, GMOs, and the Erosion of Mexican Biodiversity" in *Agriculture and Human Values* 23 (2006) and to Berghahn Journals for permission to reprint a section of "The Political Uses of Culture: Maize Production and the GM Corn Debates in Mexico" from *Focaal: European Journal of Anthropology* 48 (2006). This book expands on parts of previously published work by the Benemérita Universidad Autónoma de Puebla: my chapter in Leigh Binford's *La economía política de la migración internacional en Puebla y Veracruz* (2004); my article "¿De la economía 'natural' a la global?" in the journal *Bajo el Volcán* 7, 11 (2007); and my chapter in Francisco Gómez Carpinteiro's *Paisajes mexicanos de la reforma agraria* (2007), which was co-published with El Colegio de Michoacán and CONACYT.

Longtime friends and family members have been a source of support and encouragement throughout this project. I have counted on support and friendship from my mother Priscilla, my sister Sarah, and Steph, Sarah, and Debora. My father, Peter, deserves a special thanks for his support and editorial corrections at an earlier stage of writing. Thanks also go to my aunt Betsy for her hospitality. My newer friends Jan and Sue and Jun and Jenny have helped with making a new city of residence a home. Finally, I am especially grateful to María Mercedes Gómez for her intellectual companionship and the innumerable ways she has helped with this project.

Abbreviations

AGN General Archive of the Nation

AMC Mexican Academy of Sciences

ANEC National Association of Rural Commercialization Enterprises

APPO Popular Assembly of the Peoples of Oaxaca

BANRURAL National Rural Credit Bank

Bt bacillus thuringiensis (soil bacterium that produces insecticide)

CASIFOP Center for Social Analysis, Information and Training
(nonprofit civil association or NGO)

CBD UN Convention on Biological Diversity

CEC Commission for Environmental Cooperation (CCA in Spanish)

CECCAM Center for Studies of Rural Change

CENAMI National Support Center for Indigenous Missions

CGIAR Consultative Group on International Agriculture Research

CIBIOGEM Inter-ministerial Commission on Biosafety

CIMMYT International Corn and Wheat Institute

CINVESTAV Center for Research and Advanced Studies,
National Polytechnic Institute

CNA National Water Commission

CNBA National Agricultural Biosafety Committee

CNC National Confederation of Peasants

CNM National Maize Commission

CODEX Codex Alimentarius Commission

COMPITCH Council of Indigenous Midwives and Healers (an NGO)

CONABIO National Biodiversity Commission

CONACULTA National Council for Culture and the Arts

CONACYT National Council for Science and Technology

CONAPO National Population Council

CONASUPO National Basic Foods Company

CROM Regional Confederation of Mexican Workers

CTM Workers' Central (Union) of Mexico

DGSV   General Directorate of Plant Health

Diario Oficial   Federal Register

DICONSA   CONASUPO's distributor and importer

DNA   deoxyribonucleic acid

ETC Group Action Group on Erosion, Technology and Concentration (formerly RAFI)

EU   European Union

EZLN   Zapatista Army of National Liberation

FAO   Food and Agriculture Organization of the UN

FROC-CROC   Revolutionary Confederation of Workers and Peasants (labor union)

GATT   General Agreement on Tariffs and Trade

GDP   gross domestic product

GE   genetically engineered (also GM, or genetically modified)

GEA   Environmental Studies Group (an NGO)

GM   genetically modified (also GE, or genetically engineered)

GMO   genetically modified organism

GRAIN   an international NGO

IMF   International Monetary Fund

INAH   National Institute of Anthropology and History

INE   National Ecology Institute

INEGI   National Institute of Statistics, Geography, and Informatics

INI   National Indigenist Institute (defunct)

INIFAP   National Institute of Forestry, Agricultural, and Livestock Research

iPCR   inverse polymerase chain reaction

LBOGM   Biosafety and Genetically Modified Organisms Law

LMO   living modified organism

lps   liters per second

MAP   Mexican Agricultural Program of the Rockefeller Foundation

NAFTA   North American Free Trade Agreement

NGO   nongovernmental organization (and in Mexico, nonprofit civil association)

OEE   Oficina de Estudios Especiales

PA   Office of the Attorney General for Agrarian Affairs

PAN   National Action Party

PCM   Mexican Communist Party

PCR   polymerase chain reaction

PMS   Mexican Socialist Party

PNA   Private Newspaper Archives (of Juan Manuel Gámez Andrade, and the Barroso Archive of the Tehuacán Chronicle newspaper)

PRD   Party of the Democratic Revolution

PRI   Institutional Revolutionary Party

Procampo   direct rural support program

PROCEDE   Program for the Certification of Ejido Rights and the Titling of Urban House Plots

PROFEPA   Federal Environmental Protection Agency

PROGRESA   Education, Health, and Nutrition Program

PRONASE   National Seed Producer

PRONASOL   National Solidarity Program

PSUM   Unified Socialist Party of Mexico

PVEM   Ecological Green Party of Mexico

RAN   National Agrarian Registry

rDNA   recombinant DNA

SAGAR   former name of SAGARPA

SAGARPA   Ministry of Agriculture, Animal Husbandry, and Fisheries

SAM   Mexican Food System (policy)

SARH   former name of SAGARPA

SEDESOL   Ministry of Social Development

SEMARNAT   Ministry of the Environment and Natural Resources

SENASICA   National Service of Agriculture, Food, and Animal Health, Safety, and Quality (a branch of the Ministry of Agriculture)

SIAP   Agri-Food and Fishery Information Service

SRA   Ministry of Agrarian Reform

SUITTAR   Independent Workers Union of the Tarrant Company

TRIPS   WTO agreement on Trade Related Aspects of Intellectual Property Rights

UN   United Nations

UNAM   National Autonomous University of Mexico

UNORCA   National Union of Autonomous Peasant Organizations

UNOSJO   Union of Organizations of the Sierra Juárez of Oaxaca

WTO   World Trade Organization

THE STRUGGLE FOR
MEXICAN MAIZE

A longtime symbol of lo mexicano, or Mexicanness, maize has recently come to represent rural and even national culture threatened by neoliberal policies and corporate-led globalization in the debates about genetically modified (GM) corn.[1] Transgenes were found in local Mexican corn varieties in 2001, setting off highly charged debates about the extent to which GM corn poses a threat to native varieties in the crop's center of origin, domestication, and biodiversity. At the time the cultivation and scientific testing of GM corn were prohibited in Mexico, yet corn imports from the United States, where there is no required labeling or separation of transgenic corn, included genetically engineered varieties. Corn is imported as a grain to be used for animal feed, tortillas, and industrial processing, but it remains a seed and a living modified organism which can be planted and can reproduce in the environment. This dual nature of maize, as grain and seed, poses particular challenges for isolating or tracking GM corn in a country where native maize varieties are cultivated throughout the nation's territory. Beyond these regulatory issues, the GM corn controversy raises questions about the fate of the peasantry in an era of corporate agriculture and globalization.

In a globalized food system, foods not only travel enormous distances but have enormous regulatory, political, and cultural implications. The aim of this book is to provide readers with what one sociologist of GMOs has called a "political economy of meaning" of the corn debates, which asks under what conditions food innovations are accepted, ignored, or rejected (Murcott 2001). This book situates GM corn imports within the Mexican "neoliberal corn

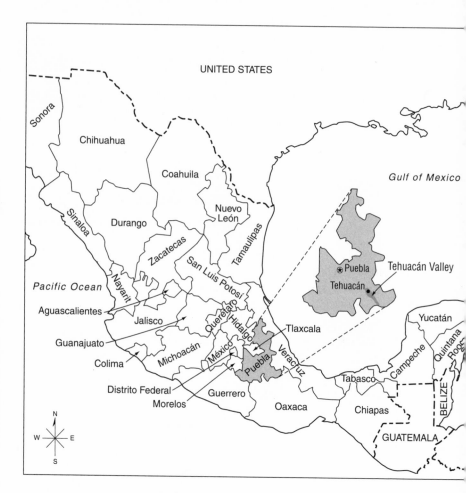

LOCATION OF THE TEHUACÁN VALLEY IN PUEBLA STATE, MEXICO

TOWNS OF THE SOUTHERN TEHUACÁN VALLEY

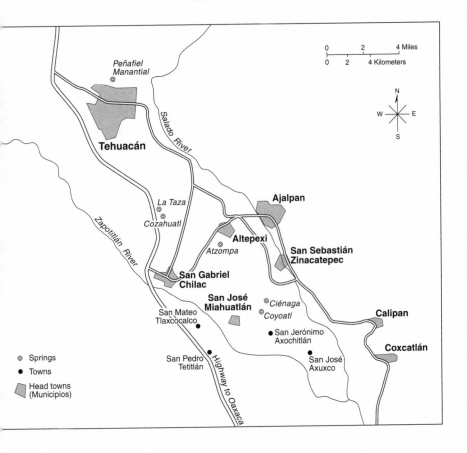

Peñafiel
Manantial

Salado River

Tehuacán

Zapotitlán River

La Taza

Cozahuatl

Ajalpan

Altepexi

Atzompa

San Sebastián
Zinacatepec

San Gabriel
Chilac

San José
Miahuatlán

Ciénaga

Coyoatl

Calipan

San Mateo
Tlaxcocalco

Coxcatlán

San Pedro
Tetitlán

Highway to Oaxaca

San Jerónimo
Axochitlán

San José
Axuxco

◎ Springs

● Towns

Head towns
(Municipios)

0   2   4 Miles
0   2   4 Kilometers

N
W — E
S

regime" (Fitting 2006a), policies which affected maize producers and consumers by bringing Mexico in line with the structural adjustment agendas of the World Bank and the International Monetary Fund. These policies advance conventional capital-intensive agriculture and the export of fruits and vegetables to Canada and the United States.[2] They promote trade liberalization, cuts to rural subsidies, and the involvement of agribusiness in various stages of production, and have deepened the country's dependency on corn imports. Mexico now imports its most consumed and culturally important crop, maize, and its most significant export is labor. In this sense neoliberal policies have sought to transform peasants into new rural subjects, into either agricultural entrepreneurs who produce for export or an inexpensive labor force. To what extent the policies have been successful in effecting this transformation is one of the questions taken up by this book.

The neoliberal corn regime also reproduces and extends older constructions of rural Mexico as a site of intervention for development. Maize agriculture, rural development, and food security are reduced to questions of profit and market efficiency. Although small-scale maize agriculturalists often have in mind reasons other than profit, or in addition to profit, when they grow maize, these reasons are devalued or dismissed. Indeed, as critics of development schemes have found elsewhere, a key component of state policy is the production of technical expertise which dismisses or excludes other types of knowledge—in this case, that of small rural producers themselves (Scott 1998; Mitchell 2002).

One of the benefits of situating the GM maize controversy within the neoliberal corn regime is that it draws our attention to how participants in the debate legitimize or challenge neoliberal policies. But when I began interviewing participants in the fall of 2000, I wondered about maize producers themselves: What did they think? How were they affected? The study of the corn debates, like the politics of food and agriculture more generally, needs to go beyond a focus on questions of regulation, policy, and state institutions to consider political practices more broadly (Lien 2004). With these concerns in mind, I carried out fieldwork on the livelihood strategies of smallholder maize producers and residents in the Tehuacán Valley of Puebla state. These producers constitute an important layer of meaning in the recent GM corn debates, not just because their practices and voices are the subject of debate but because they are actors who react to and engage rural policy and state bureaucrats and experts. The valley is also one of the sites where evidence of transgenes was later found in native cornfields.[3]

While the Tehuacán Valley differs from communities politically active on the issue of GM corn, such as its neighbors in the highlands of Oaxaca, like much of rural Mexico the valley is struggling with neoliberal policies and crisis. The valley represents a common and significant disjuncture in the maize debates: what is under debate in the Mexican Senate, in national newspapers, at academic conferences, and at urban rallies may not be a topic of conversation or debate in the countryside, and when it is, the debate is often framed in distinctive ways. The information and debate about GM corn is unevenly communicated, shared, and received. As GM corn and the controversy surrounding it move from one social context to the next, they are translated and understood in different ways.

This book examines the livelihood struggles of maize producers in relation to the questions and issues raised by the GM corn debates. The future of in situ maize conservation depends on the regulation of GM imports, but perhaps more importantly on the livelihood practices of rural Mexicans. Maize biological diversity is affected by the social relations of production and reproduction among growers. It is a dynamic process in which native maize varieties (criollos) are maintained and developed through exchanges between cultivators and between cornfields. As the Mexican critics of transgenic maize and their allies in the transnational food sovereignty movement have pointed out, if the small-scale producers who select and plant regionally varied types of maize abandon agriculture in large numbers, the in situ genetic variety and abundance of the crop will be displaced. I do not mean to imply that change in rural livelihoods necessarily has negative social or ecological consequences. Rather, one of my main arguments is that although small-scale agricultural producers are always faced with a degree of uncertainty, under the neoliberal corn regime the struggle to maintain or improve their livelihoods has intensified. Moreover, I believe that those campesinos who want to remain on the land should have the ability to do so. The anthropologist Armando Bartra (2008) refers to this as the right not to migrate, "the right to stay home."

This introduction briefly summarizes the effects of the neoliberal corn regime on the southern Tehuacán Valley, and then outlines a second argument of this book: that the debates over GM corn—how the issues are framed, what is said and not said, and by whom—have significant political consequences. Some debate participants would benefit from listening to the perspectives of small-scale maize farmers on the difficulties that they and their communities face. This introduction also provides some background on economic crisis and structural adjustment policies in Mexico, discusses my methods and theoretical approach, and takes up key concepts such as peasants, food regimes, neoliberalism, and globalization.

OVERVIEW OF THE BOOK

Part I of this book, "Debates," examines how questions of culture, risk, and expertise are framed in the controversy surrounding transgenic maize. Based on interviews with participants in the GM corn debates, attendance at coalition forums and press conferences, and the reading of media reports, chapter 1 examines how the official government position relies on scientific experts to evaluate the risks associated with gene flow among maize varieties, while the anti-GM corn coalition calls for including Mexican campesinos, consumers, and concerned citizens in the risk assessment process. Chapter 2 focuses on how particular assumptions about rural culture in early agrarian policy and debate are used to contest or defend more recent neoliberal policies. The anti-GM coalition challenges the official perspective by drawing on the Mexican and transnational food and peasant rights movements, which highlight the goal of food sovereignty.

Part II of the book, "Livelihoods," considers the practices and perspectives of southern valley maize producers, migrants, and maquiladora workers. It begins by taking readers into the Tehuacán Valley to consider how an indigenous peasantry was formed in interaction with the practices of the state and capital accumulation. Chapter 3 provides a snapshot of the valley town San José Miahuatlán at mid-twentieth century to illustrate the centrality of irrigation water in maize agriculture and community organization. It explores local disputes over water and ideas about indigenous ethnicity. Chapter 4 argues that households have dealt with economic and environmental crises by combining maize production with maquiladora work and migration to the United States, and that this in turn has led to the emergence of transnational peasant households and households in transition. While migrant remittances help to support valley households, these funds and the experience of waged, off-farm employment are changing ideas about the generational and gendered labor of the household, including maize agriculture. Notably, as explored in chapter 5, migrants in their teens and twenties have little knowledge of agriculture, or experience or interest in agricultural production. This last chapter also takes a closer look at local narratives about corn agriculture and the reasons why a campesino identity resonates with older residents.

RESEARCH IN THE CRADLE OF CORN

Located in the southeastern end of Puebla, the Tehuacán Valley descends from north to southeast, continuing toward Teotitlán, Oaxaca. The Sierra Zongólica mountain range, which forms part of the Sierra Madre, borders the valley at its northern and eastern sides. The valley is also bordered by the Sierra de Zapotitlán and the Sierra de Mazateca at its southern and western sides. In 1998 the Tehuacán-Cuicatlán Biosphere in southeastern Puebla and northeastern Oaxaca was established, to help protect the biodiversity of cactus and other species in the region from the threats of deforestation, overgrazing, and illegal sales.

The valley is known as the "cradle of corn" because of Richard MacNeish's important archaeological study of the 1960s, which uncovered maize cobs dating back to 5000 BCE (MacNeish 1972). Although the valley is considered one of several possible locations of original maize domestication, recent evidence suggests that other sites are more likely candidates (Matsuoka, Vigouroux, Goodman, Sanchez, Buckler, and Doebley 2002).

Prehistoric irrigation was central to this incipient agriculture (Woodbury and Neely 1972), and even today, an irrigation system of water springs, underground tunnels, and chain wells (*galerías filtrantes*) remains essential to agricultural production in the valley. In the late 1920s water bottling plants were established in Tehuacán, and soon after, the city attracted tourists to its spring waters, believed to have healing properties.

Today campesinos and indigenous peoples from the valley and surrounding sierras look to the growing city of Tehuacán for employment in spring water and soda bottling plants, the poultry in-

dustry, and apparel plants, or maquiladoras. The region was nicknamed the "capital of blue jeans" during a maquiladora boom in the
1990s. The area has a mixed heritage of Nahua, Popoloca, Mixteca,
Chocho, and Mazateca peoples, although Nahuatl became the common language of the valley through Aztec domination shortly before
the Spanish conquest (Aguirre Beltrán 1992 [1986]). Nahuatl is the
most widely spoken indigenous language in Mexico.

Maize and beans are commonly cultivated crops in the valley, and
commercial crops like garlic, tomatoes, sugarcane, fruits, and flow-
ers are also grown. As in other areas of Mexico, rain-fed white
corn—distinct from industrial, hybrid yellow corn—is largely grown
for human consumption. Since the 1960s valley producers have also
grown irrigated white maize for sale on the cob, called *elote*. Other
significant activities in the region include goat herding, the production of construction materials (especially bricks and cinder blocks),
and handicrafts like baskets and embroidered clothing for tourist
markets outside the valley.

South of the city there are seven valley *municipios*, or counties,
which cultivate commercial elote. San José Miahuatlán (pop. 13,500)
is the southern municipio bordering Oaxaca, comprising the head
town (*cabecera*) of the same name and four auxiliary towns: Axusco,
San Jerónimo Axochitlán, San Pedro Tetitlán, and San Mateo Tlacox-
calco. The population of the head town, where my research was
focused, is around 8,760.[4] While state authorities classify the county
of San José as a "marginalized" indigenous area because it is one of
the poorest areas of the valley (Embriz ed. 1993, 159–60), it is also
considerably better off than the neighboring sierra in terms of services like potable water, electricity, and transportation.

In the 1980s the anthropologists Kjell Enge and Scott Whiteford
found that agriculture was "the lifeblood of the Tehuacán Valley"
(1989, 29). This holds true today, although livelihoods have further
diversified and migration has accelerated. In San José cornfields
(*milpas*) either follow the Mesoamerican tradition of intercropped
maize, beans, and squash or are simply limited to corn. Landholdings tend to be small (up to five hectares) or less frequently of
medium size (six to twelve hectares). These holdings are on communal, private, or *ejido* land, the last consisting of hillside terrains
largely used for wood collection, goat grazing, and to a lesser extent
rain-fed maize production. Rainfall in the valley is irregular, and

while the soil is fertile in many areas, calcium salts and carbonates are deposited in the soil by irrigation water. Over time this can lead to salinization and soil that becomes toxic to plants. Additionally, when the concentration of calcium salts is high, they form a hardpan beneath the surface (*caliche*), making drainage of the soil very difficult (Byers 1967; Enge and Whiteford 1989, 27–28).

Methodology

This book focuses on the debate over the GM corn scandal during a six-year period corresponding to the administration of President Fox (2000–2006). I interviewed various types of debate participants about regulation and the GM corn controversy—government officials, maize biologists, biotechnologists, and anti-GM corn activists in Mexico City, Tehuacán, and Chapingo—who were identified as experts in the media or by other participants. I wanted to understand how they discussed and framed the controversy and the problems facing the countryside. Social scientists are increasingly interested in the role of experts and expert knowledge in state practices and political rule. Modern states and public officials often rely on the "rule of experts" (Mitchell 2002), as expertise enables them to present their decisions in technical rather than political terms (Ferguson 1994). In Mexico the study of "experts" and those with influence includes looking at how anthropologists and social scientists portray the countryside; over the years anthropology has played an important role in shaping Mexican state policy and representing rural folk. This role is discussed in chapter 2. By critically engaging our own discipline, anthropologists can be more conscious of the ways we contribute to the construction of rural Mexico as a site for particular types of expert interventions.

The second method I undertook was ethnographic fieldwork, which I conducted among residents in the southern Tehuacán Valley town of San José Miahuatlán. I lived in the valley in 2001–2, with several extended visits over the next six years. When fieldwork began I asked residents what they thought about *maíz transgénico*, but found that the controversy had not reached the valley during my visits, despite the government study which found evidence of transgenes in its northern end. I carried out research with a couple of other questions in mind: Why was maize the crop of choice when local agricultural production declined in the 1980s and into the 1990s—as residents and statistics suggested? And how were livelihoods affected by trade liberalization and recent state policies?

Fieldwork included seventy interviews and surveys with residents on their household composition, work or migrant history, and agricultural practices and corn varieties. Over a period of several months in 2002 I also accompanied a government extension worker from the regional office of the Ministry of Agriculture during his visits to producers in valley towns. Together we conducted surveys on the costs of corn production.

For anthropologists, fieldwork is based on participant observation, which is much more than conducting interviews or surveys; it is a process of building relationships with residents and communicating informally with people in everyday situations, such as when they are hanging out at the corner store, at a friend's home shelling corn, or in line at the mayor's office waiting for some subsidy. Of course anthropologists are not neutral or disinterested observers: they occupy particular social locations and take with them to the field questions that have been shaped by their academic training and life history. My social location as a university-educated North American gave me access to government offices and research sites that less privileged rural Mexicans do not have. Moreover, my account of the valley is not a complete picture of life in the region. Ethnographic fieldwork does not get us to the "truth" or total picture of a place, but it does provide a rich context for understanding interview or survey responses and does give us insight into ideas and practices that may not be captured in other ways.

When I first arrived in San José Miahuatlán I quickly found out that tensions existed between the local branch of the PRI (Partido de la Revolución Institucional or Institutional Revolutionary Party)[5] and the other main political party in town, the PRD (Partido de la Revolución Democrática or Party of the Democratic Revolution),[6] because of conflict over irrigation water during the 1980s. At the national level the PRI had been in power for seventy-one years. I made an effort to associate with families of different political affiliations and took special care to change the names of the interviewees in my notebooks and in publications, with a few exceptions.[7] History weighs heavily in San José. Although many Mexican rural communities, including other valley towns, have experienced periods of violent conflict over scarce resources, in San José the conflict has shaped the relations of community in particular and profound ways. There is a perception among valley residents and city dwellers that Sanjosepeños are naturally prone to violence.

San José has both similarities with other struggling rural towns and its own unique history. Like other rural areas of Mexico it has a history of postrevolutionary disputes over resources and now combines maize production with transnational labor migration. However, in the valley producers grow both rain-fed and irrigated corn, in contrast to peasant producers who depend on rainfall alone. Moreover, in a country so regionally varied by language, custom, and geography from one community to the next, the kind of maize produced and the labor strategies employed vary greatly, as can the specific reasons for conflict and labor migration, and their effects.

Valley Livelihoods: Maize, Migrants, and Maquiladoras

When I began fieldwork in the southern valley I found that many households cultivated corn for consumption and sale on the market; and that they were financed by off-farm income, including remittances from young migrants in the United States and employment in valley maquiladoras. Yet in post-NAFTA Mexico it is often less costly to buy industrially produced and imported corn than it is to grow the crop locally on a small scale.

Previously not a migrant-sending area, San José now sends the majority of its young male residents (aged fifteen to late thirties) to work across the border. At home maize is the preferred crop because it is the mainstay of the diet and has multiple, flexible uses: if there are no buyers or the price of maize is too low, the crop is dried and consumed as grain in the form of tortillas. When cash is needed in emergencies, the grain can be sold in small amounts at a loss. In the absence of a social safety net in rural Mexico, maize provides a kind of insurance, particularly for older residents and the unemployed. Maize is a form of security for those left behind, for those who cannot migrate or do not wish to. In other regions maize agriculture, cuisine, and seed can also have a strong spiritual component.

In the southern Tehuacán Valley residents are struggling with the effects of inflation, lack of regional employment, and neoliberal policies, as well as with declining levels of irrigation water. As a result, agriculture and the social relations of production are being remade in significant ways. The household strategy which combines maize production with migration and maquiladora work has been accompanied by the monetization of available agricultural labor and a decline in sharecropping. Young men hired to work the milpa now prefer to be paid in wages rather than through sharecropping arrangements. (In the valley, work in the milpa is typically done by men, although women contribute to other aspects of maize agriculture.) More significant is the preference of young migrants for non-agricultural work. Members of this younger generation have little knowledge about corn agriculture and claim that they will not take up the crop as they age because "there is no money to be made in the cornfield." A last trend is the declining use of the traditionally inter-cropped milpa (maize grown with beans and squash) and several varieties of local maize.

NARRATIVES ABOUT CORN CULTURE

The politics of food and agriculture involve struggles over who directs the focus of public debate and how the issues are framed. In Mexico official narratives articulated by the Ministry of Agriculture and in policy portray smallholder corn agriculture as inefficient because of its low yields and its use of "traditional" technology, such as criollo seed. Drawing upon interviews with Mexican scientists and activists engaged in the corn debates, this book demonstrates how the coalition In Defense of Maize, formed in 2002 by Mexican environmentalist, campesino, and indigenous rights groups in response to transgenic contamination, shifted the debate away from the official focus on inefficiency and the risks of gene flow toward wider concerns about the future of the Mexican countryside and culture. Critics of GM maize challenge the government and industry narratives which privilege scientific expertise in evaluating the risks of gene flow (see Heller 2002 on France). In doing so this "pro-maize", anti-GM coalition contends that the appropriate experts for evaluating potential harm are not only biotechnologists and other scientists but corn producers and consumers. As with other GM controversies, the Mexican corn debates contest the boundaries of accepted expert knowledge and also implicate competing constructions of culture and nature. The anthropologists Chaia Heller and Arturo Escobar further suggest that in such controversies, "Bio-diversity and transgenic agriculture constitute powerful networks through which concepts, policies and ultimately cultures and ecologies are contested and negotiated" (2003, 169).

Participants in the GM corn debates articulate ideas about peasants, indigenous peoples, and development in relation to corn and culture. Arguments about culture are used to defend or reject recent state policies and trade liberalization. While the pro-maize coalition draws our attention to the policies which exacerbate the difficult conditions for small producers, in some cases they misrepresent changes taking place in the countryside. At times both the government and the pro-maize coalition portray maize agriculture as part of a millennial culture or tradition, distinct from the capitalist economy of modern Mexico—a form of what Michael Kearney (1996) and others have called "peasant essentialism."[8] The government narrative posits the production of corn as inefficient precisely because it is deemed a culture of subsistence untouched by the workings and values of capitalist markets. Critics counter that this is a positive alternative to capitalism and its processes of commodification. Both these narrative strategies rely on a conceptual binary between the "market" and the "local community" (Hayden 2003b) and between the modern and the traditional. They also overlook the fact that Mexico, and the rest of Mesoamerica, were an important pre-capitalist center of commodity production and exchange (Cook 2006). Thus an additional argument of this book is that conceiving of a millennial culture of corn obscures how maize-producing communities (or peasantries) are made and remade in interaction with larger forces and processes.

These narratives about the Mexican countryside are not simply words and ideas; they are an inherent part of social practice and have material effects in the world. There is power in the process of naming. The ways that policy makers and experts view and describe the countryside, its problems and remedies, make their way into policy and state practices, although these policies are implemented and received in uneven and unintended ways. Various social constructionist schools of thought rightly point out that we never arrive at the truth about the social world in a manner unmediated by language, discourse, or ideology; nevertheless, some representations are fairer or more rigorous than others. This book is of course my own account of how policies and narratives make their way to the countryside. It is my hope that this ethnography illustrates why we should not rely on official versions about the benefits of trade liberalization, conventional agriculture, measures to cut state services, or neoliberal solutions to rural poverty. Alternative accounts of these policies show that campesinos are not "inefficient" producers, isolated from the workings of the market, nor have they necessarily responded to neoliberal reform in the predicted manner. Valley migrants, campesinos, and maquiladora workers are social agents who engage and respond to state policies and globalization, but not under conditions of their own choosing.

Conceptualizing Culture

As both a crop and a food for humans, mainly in the form of tortillas, maize is a particularly powerful symbol of the nation in Mexico, with many often contradictory layers of meaning. Foods have strong emotive powers because they structure daily life, are part of the process of socialization, and are symbols and signs of other things. The act of eating involves consuming meaning and symbols as well as consuming foods (Douglas 1966; Mintz 1985). Foods play a role in demonstrating and delineating social distinctions such as social status, ethnic belonging or exclusion, and gender difference. In Mexico corn-based foods are inscribed with notions of culture, race, and gender, and so is maize agriculture. In chapter 2 we see how maize-based foods went from being a symbol of indigenous backwardness and isolation to a symbol of the mestizo nation in mid-twentieth-century Mexico, yet state policy continued to associate maize agriculture with economic backwardness and inefficiency. And while areas that rely most heavily on criollo varieties (rather than scientifically improved seed) do tend to be the poorest in Mexico, it is sometimes incorrectly suggested that small-scale maize production is responsible for rural poverty.

The recent GM corn debates have inherited and negotiated earlier ideas about peasant maize-based agriculture and rural culture: they are frequently portrayed as isolated, primordial, and driven by values of subsistence over profit. Such peasant essentialism distorts changes taking place in the countryside and the strategies of smallholder maize farmers and campesinos as they confront the neoliberal corn regime. This portrayal overlaps with a bounded and internally static concept of culture—what the anthropologist Eric Wolf (1982) famously referred to as the "billiard ball" view of culture. In recent years anthropology has criticized such bounded views of culture, though perhaps overemphasizing the delinking of culture from particular places through processes such as transnational migration (Escobar 2001). Although place is socially and historically constructed, "place-based practices and modes of consciousness [still matter] for the production of culture" (ibid. 147). Indeed the southern valley town of San José is very much a transnational community these days, as residents' ideas and experiences of being Sanjosepeño are influenced by migration; however, this does not change the need to understand how their sense of community and identity are also place-based practices, shaped through the history of the valley.

In the hopes of going beyond the stereotypes about campesinos and maize-based rural culture, be they romantic or disparaging, this book draws on an anthropological tradition which foregrounds history and power and insists on considering both the structural features of capitalist globalization and their historical and geographic contingencies. This book does not treat peasant livelihoods and maize agriculture as a millennial tradition or a contained cultural logic or system, even though peasants' agricultural knowledge about seed varieties and cultivation can be systematic (see chapter 5). Rather, this book approaches maize agriculture as "culture": unevenly shared meanings and practices which are formed in engagement with wider economic and political structures and social class formation. Culture is understood in historical and dynamic terms; as meanings, practices, and relations which have a selective continuity with previous generations. In an anthropological sense, culture refers to a way of life including social conventions, institutions, and forms of production, but also to what Raymond Williams (1961) has called a "structure of feeling": an ongoing, active process of cultural production in which aspects of lived experience are the raw material for alternative and perhaps even oppositional values.

In the southern Tehuacán Valley residents share some ideas and practices about maize agriculture, although they tend to differ by generation. These ideas and practices emerge in interaction with wider economic and political processes, and should not be viewed as previously contained or isolated entities ("cultures" or "natural economies") that now, for the first time, confront change and the outside world, namely the global economy.

Corn in Context

In English "maize" and "corn" are generally used interchangeably. The word "corn" once referred to all cereals, with "Indian corn" referring specifically to maize (*zea mays*), the crop from the Americas. Eventually this name was shortened to just "corn" (Fussell 1992). In Mexico maize is cultivated in a variety of ways, from industrial production to subsistence farming, and in a vast array of environments. It is a central part of rural livelihood strategies. An estimated three million people work directly in corn production, upon which 12.5 million rural inhabitants depend. The crop occupies half of Mexico's cultivated land (SAGARPA-SIAP 2004).

Maize is also the traditional staple of the Mexican diet, particularly in the form of tortillas. Corn-based food makes its way into Mexican diets in three main ways. First, the maize producers themselves consume maize in the form of tortillas, *atole* (a hot drink), tamales, and other foods made from white, blue, red, and yellow maize varieties. Traditionally, tortillas are made through the process of nixtamalization, or soaking the corn in water and mineral lime, (calcium carbonate), causing the skin of the kernel to peel off. This process also releases the vitamin niacin and the amino acid tryptophan in the corn. The corn is then ground and kneaded into dough (*masa*), patted out by hand or with a small press, and cooked on a *comal*. Maize growers often sell a small part of their harvest to local markets or distributors as grain or fresh corn to be eaten at home or prepared by street vendors as corn on the cob (elote). In Mexico every part of the corn plant is used by smallholder farmers: the husks are dried and used to wrap tamales, the shelled cobs are burned as fuel for fire, and the leftover stalks are given to animals as feed.

The second way Mexicans acquire tortillas is from corn mills and tortilla sellers. The mills (*nixtamaleros* or *molinos*) are paid to grind the nixtamal into tortilla dough, which is then made into tortillas at the store or at home. Traditionally tortilla stores and sellers (*tortillerías*) have relied on nixtamalized dough made from white maize. However, tortillas in urban Mexico are increasingly made from ready-mix corn flour that becomes dough when water is added. Although raw corn milled into flour has a shelf life of up to three months compared to the one- or two-day shelf life of nixtamal corn, it does not have the nutritional benefits of nixtamalization.

The third way that maize is produced for food is through the corn

TABLE 1: Estimated Demand for Corn (Grain) in Mexico

Uses of Corn in Mexico: Annual Consumption (in million tons)

	2004	2005	2006
WHITE CORN			
Flour	3.5	3.2	3.7
Traditional tortilla industry	3.3	3.0	3.4
Tortilla consumption in rural areas	3.4	3.1	3.5
Total human consumption	10.2	9.4*	10.6
Animal consumption	2.1	1.9	2.2
Total white corn	12.3	11.3	12.8
YELLOW CORN			
Cereals and snacks	0.5	0.4	0.5
Livestock feed sector	11.8	10.8	12.3
Other uses	2.6	2.4	2.7
Total yellow corn	14.7*	13.6	15.3*
Total	27.0	24.9	28.2*

* Thus in original: sum of individual items differs from total (rounding error cannot account for discrepancy).

Source: "Situación actual y perspectivas del maíz en México, 1996–2012" (Servicio de Información Agroalimentaria y Pesquera (SIAP), n.d.).

flour and tortilla industry (Nadal 1999, 119). This industry sells corn flour as a ready-made tortilla mix to tortillerías (including those in supermarkets), along with prepared and packaged tortillas to grocery stores and supermarkets. The value of this corn flour industry is close to $4 billion, while the corn-based snack industry in Mexico is worth over $1 billion (SAGARPA-SIAP 2006). This industry is concentrated in the hands of the Mexican corporations Gruma, which includes the well-known subsidiary Maseca, and Minsa, and Cargill the transnational corporation based in the United States. The majority of corn destined for corn flour, starch, and cereal producers, or to be used as animal feed—which then is consumed by humans as meat—is yellow maize imported from the United States or bought from medium- or large-scale Mexican farmers. Most imported yellow corn is a transgenic variety. Since maize was one of the few crops under NAFTA to retain some subsidy for producers (under the Procampo program), medium-sized farmers, based largely in northwestern Mexico, took up corn for commercial production on irrigated land. This new group of entrepreneurial corn farmers was able to profit by using high-yielding and largely non-GM varieties— although in 2008 illegally cultivated transgenic maize was found growing in Chihuahua.[9]

Although Mexico produces enough white corn for domestic food consumption, with the growth of the tortilla and corn flour industry, corn-based foods are increasingly made from yellow corn. The industry places Mexico's small-scale white corn growers at a disadvantage: when the big corporations do not like the price of local corn, they buy imports. This puts pressure on the price. This also means that this industry has taken over a portion of the market that previously belonged to domestic white maize producers.[10] In other words, the pro-industry policies of recent years have undermined Mexico's self-sufficiency in maize for domestic food consumption. Such policies in Mexico reflect an expansion of agribusiness and neoliberal policies on the international level, or what some authors have characterized as an emergent international food regime.

Food from Nowhere: An Emergent Corporate Food Regime

The concept of a "food regime" was first used by Harriet Friedmann (1987) to describe policies, norms, institutions, and trade relations related to food and agriculture between unequal nations.[11] The first international food regime emerged in the late nineteenth century, when Britain was at the center of key food circuits. Tropical foods were imported from the colonies and grains and meats from the settler colonies (McMichael 2009, 144). After the Second World War a second food regime developed, as the United States used food aid to create new markets in the global south for its grain surpluses (Friedmann 1987). This food aid encouraged dietary shifts, including the consumption of grain-fed livestock and processed foods. Although this regime promoted conventional agriculture and Green Revolution technology in the name of national development, in practice international agribusiness expanded and national farm sectors were undermined (ibid.; McMichael 2009, 146). Mexican state policy was influenced by this international regime in particular ways, as discussed in chapter 2.

A third, emergent regime is characterized by the expansion of corporate agriculture and power in world institutions like the World Trade Organization (WTO), the continued export of subsidized grain from the global north, and the rise of nontraditional food exports (fruits, vegetables, and meats) from the global south, produced for agribusiness. As Philip McMichael points out, the WTO and trade agreements like NAFTA preserve farm subsidies in the global north, "while Southern states have been forced to reduce agricultural protections and import staple, and export high-value, foods" (2009, 148).

This globalization of corporate agriculture expands and deepens capitalist relations. In the process it often generates "populations of displaced slum-dwellers as small farmers leave the land" (2009, 142). Additionally, the corporate agro-food system involves further dietary shifts, as more consumers from around the world eat processed foods found in supermarkets. This agro-food complex produces what the French farmer, activist, and Confédération Paysanne leader José Bové has called "food from nowhere," or food purged of "taste, health and cultural and geographical identity" (cited in Desmarais 2007, 28). Industrial tortillas in Mexico embody the corporate food regime's "food from nowhere" in the above senses. The neoliberal corn regime is a concept which refers to the particular configuration of this international food regime in Mexico.

Pechlaner and Otero (2008) add that this regime is characterized by the rise of genetic engineering as the main technology for capitalist agriculture and by changes in regulation which accommodate this technology. However, they importantly point out that "despite prevailing trends, sufficient local resistance to the technology could modify or even derail, the technology's role in individual nations, and accordingly, in the unfolding food regime as a whole" (352). In Mexico the coalition In Defense of Maize acts as a policy watchdog and challenges industry claims. The activities of the coalition, along with the livelihood struggles of small-scale producers, are a reminder that food regimes do not encapsulate all food-related practices; there are other production and consumption practices which contradict, resist, or negotiate and modify the norms and goals of food regimes (McMichael 2009, 146).

CRISIS IN THE COUNTRYSIDE

After a drop in oil prices an economic crisis hit Mexico in August 1982, and the government of López-Portillo (1976–82) announced that it could not meet its debt payments. Peso devaluations, inflation, and debt renegotiations followed (1982, 1989, 1994–95). The countryside was hit hard. By the late 1990s the United Nations Commission for Latin America and the Caribbean characterized the situation in the Mexican countryside as "an authentic crisis." López-Portillo's successor, President Miguel de la Madrid (1982–88), quickly implemented austerity measures, and the International Monetary Fund (IMF) prescribed stabilization programs. During his six-year term, or *sexenio*, the government privatized 743 state enterprises and reduced state expenditures from 30 percent of GDP (in 1981) to 17 percent. Real wages dropped about 60 percent. Carlos Salinas de Gortari was appointed minister of budget and planning; and six years later, as president (1988–94), he implemented a radical reform of the agrarian bureaucracy and negotiated the North American Free Trade Agreement, or NAFTA.

The country's economic problems were followed by a number of political crises that rocked the already precarious legitimacy of the ruling party, the PRI (which had been in power since 1929), and that forged a more democratic path for the country. There were five key moments of political crisis: the government's inadequate response to the Mexico City earthquake of 1985, the fraudulent presidential elections of 1988, the EZLN uprising in the southern state of Chiapas in 1994, the loss of the PRI majority at the national level in 1997, and the victory of the opposition party, PAN, in late 2000.[12] With the victory of the PAN, Vicente Fox, a former director of Coca Cola Mexico, became president (2000–2006). More recently the PAN's legitimacy was questioned when two political crises unfolded during the last year of President Fox's tenure: the contested electoral win of President Fox's successor, Felipe Calderón, also of the PAN, and the violent repression of the Popular Assembly of the Peoples of Oaxaca (APPO), which grew out of a teachers' strike and encampment in downtown Oaxaca city in May 2006.

Following neoliberal economic doctrine which gained favor internationally in the late 1970s and 1980s, the IMF and the World Bank prescribed economic austerity programs, the privatization of state services, and a new round of market-oriented intervention. By the end of the 1980s official development policy was promoting the allocation of resources through the market rather than the state. In Mexico the administrations of Salinas and his predecessors took a "shape up or get out" approach to smallholder farmers, advising them to adapt to market liberalization, the privatization of state enterprises, and cutbacks to funding, or to find a different occupation. Economic crisis coupled with these policy shifts induced rural Mexicans to seek out a livelihood elsewhere.

Mexican laborers had worked in the United States throughout the twentieth century, but in recent years rural Mexicans have migrated to cities in the United States (and to a lesser extent Canada) in record numbers. The remittances that they and their compatriots sent home reached $25 billion in 2008—with a drop after the economic decline of the following year.[13] The significance of rural migration, however, is interpreted in different ways.

THE TWILIGHT OF THE PEASANTRY?

The debate about rural migration is related to much older debates about the fate of the peasantry under capitalism. In the early twentieth century, where V. I. Lenin saw the inevitable proletarianization of the peasantry, A. V. Chayanov saw a peasant logic of non-accumulation which withstood the disintegrative forces of capitalism. The 1960s and 1970s saw a renewed debate about peasantries, influenced by the Vietnam War and the liberation movements of the largely rural countries of the global south. In Mexico research was often polarized between the agrarian populists, the "peasantists," who focused on the function of peasants as cheap, reserve labor for capitalists, and the "proletarianists" who focused on an inevitable or desirable transition to waged work. There were important feminist interventions in this debate on the gendered work of social reproduction, and later on whether employment (say, in export processing zones and maquiladoras) frees young women in patriarchal rural societies from the tutelage of their male relatives or rather generates other forms of gendered discipline and exploitation (Ong 1987).

Peasants were conceptualized as smallholder producers who farm land (owned or rented) with their own labor and the unpaid labor of family or sharecroppers to provide for the consumption needs of their households and sometimes for exchange or sale. Peasant community relations were seen as embodying cultural norms and moral expectations of reciprocity—a moral economy (Scott 1976). These norms were often a form of risk avoidance, in that the subsistence needs of the household took priority over efforts to maximize agricultural returns. Peasants were seen as exploited and dominated through unequal exchange and the extraction of rents and taxes. In this way the agrarian debates considered questions of power and the social relations of production, but sometimes fell back into an essentialist conception of the peasantry or into a teleological definition of capitalism (which among other things precluded the possibility of peasant agency). Peasants were sometimes treated as a homogeneous category, in a manner that overlooked social and economic differences, or treated the process of differentiation as the result of externally imposed capitalism (Cook and Binford 1990, 18). An additional problem is that although the practices and expectations of reciprocity and risk avoidance do characterize many peasantries, peasant communities have been romanticized as being predisposed to simple reproduction, averse to profit, or constituted by egalitarian relations. Another pitfall of the agrarian debates was that domination and exploitation were sometimes viewed as external to peasant relations, not affecting their practices or norms of reciprocity. In fact, the term "peasant" became so encumbered with assumptions and political baggage that some scholars stopped using it altogether.

There were, however, participants in the agrarian debates who avoided these pitfalls and considered peasants agents and actors and not just the victims or instruments of structural or systemic conditions. Some emphasized that agrarian capital could emerge as an endogenous development (Cook and Binford 1990), or that peasants are not isolated from global capital but have long been influenced by capitalist relations of production and reproduction (Roseberry 1993).

The debate about the fate of peasants continues today. With changes to global capitalism since the 1970s, peasantries are now more connected and therefore vulnerable to the vagaries of the world market. But are we witness to, as Eric Hobsbawm has suggested (1994, 289), "the twilight of the peasantry"? Most of the world, after all, now lives in urban environments. The way this question is answered depends in part on how the peasantry is defined.

In this book I use the term "peasants" or "campesinos" to refer to small and medium-sized producers who combine agriculture for their own consumption with some production for sale, but who may also rely on off-farm employment. Since peasant households must purchase those goods which they do not produce themselves, they need to generate cash either by selling their crops or crafts (as petty commodity producers), engaging in small commerce, or working for wages. Of course this expanded definition of the peasantry has a threshold at which it no longer holds, such as when the farm labor and other agricultural inputs (seed, fertilizer, irrigation, etc.) are monetized or when a rural household has abandoned farming and relies completely on waged employment. But what about cases that are less clear-cut? What about rural households that combine un-paid smallholder agriculture with paid agricultural labor and off-farm employment yet self-identify as campesinos, as is often the case in the southern Tehuacán Valley? Drawing on some of the insights from the earlier agrarian debates, an expanded definition includes such cases but prompts us to ask whether, and to what extent, agriculture is based on waged or unwaged labor, and to ask *why* the term "campesino" is used as a self-label. My definition is thus meant to reflect the diversification and complexity of peasant livelihood practices in rural Mexico, but also the fact that the term "campesino" is a political identity in Mexico and much of Latin America (Edelman 1999; Gledhill 1985).

Although campesinos are often portrayed as belonging to a static and homogeneous category, their identity emerged in Mexico as a highly politicized and resonant one during the revolution of 1910–20 (Boyer 2003). The revolution heralded campesinos not only as the rightful owners of the land but as the heart and soul of the nation. In the early twentieth century in places like the Tehuacán Valley where residents worked as agricultural day laborers on land they once owned (and with irrigation water they once controlled), this image of the campesino resonated. So while revolutionary lead-ers and the post-revolutionary state employed the symbol and iden-tity of the campesino to garner political support for their state-building project, at the same time many rural peoples identified with the image of the campesino based on their experiences of disen-franchisement and collective memory of injustice (Boyer 2003). The campesino identity was used to negotiate and challenge the state. This was true of the southern Tehuacán valley, where residents em-ployed a notion of campesino, alongside or in contrast to an indige-nous identity, in struggles over land and water.

While to some scholars rural migration in Mexico is evidence of the displacement of the peasantry and the proletarianization of the countryside (Kearney 1996), others contend that migrant remittances help to renew peasant production or create a new rurality (Barkin 2002). Scholars also disagree about whether migrant remittances help raise the rural standard of living and encourage economic development or rather signal class differentiation, raise consumption expectations, and do little to alter rural dependency on the United States labor market (Binford 2003; Cohen, Jones, and Conway 2005; Delgado Wise 2004; Durand, Parrado, and Massey 1996; Jones 1992).

An important contributor to Mexican scholarship, David Barkin, argues that peasants defend their maize-based livelihoods through remittances and the marketing of peasant-made goods, like handmade tortillas; and that this selective market engagement strengthens rural society, constituting an alternative to neoliberal globalization (Barkin 2002; Barkin 2006). "New rurality" studies helpfully point to the increasing diversification of agriculture and its reliance on remittances, the flexibilization of the rural workforce, and the growth of agribusiness in agriculture; trends which are all found in the Tehuacán Valley. However, this approach can also overlook the role of the state and class relations—particularly social and economic differentiation in the countryside—or paint too positive a picture of recent rural change (Kay 2008).

To return to the question about the future of the peasantry, crisis and neoliberal policies have displaced many rural Mexicans, including half a million agricultural workers between 1995 and 2005 (Pérez, Schlesinger, and Wise 2008). Yet in certain regions like the Tehuacán Valley, these processes have not brought about an end to the peasantry—at least not yet—but have entailed the remaking and further diversification of rural livelihoods. In some cases the category or self-label of campesino may obscure the fact that such producers are no longer peasants but rural proletarians or commercial farmers, or in a process of transition from one to the other. In other cases the self-label of campesino may be a renewed or reformulated political identity for collective action against neoliberal policies, as with some participants in the coalition In Defense of Maize or the transnational organization Via Campesina.

In the southern Tehuacán Valley household survival depends on combining migrant wage labor and other income-earning strategies with small corn production for exchange, sale, or household consumption. This reliance on nonagricultural income is not new, but it has intensified. Throughout the twentieth century residents combined agricultural production with waged labor, and sold goods in regional markets; but more recently, this paid employment has generally been found in valley agribusiness and maquiladoras, or in the United States.

Today older residents self-label as campesinos and rely on the migrant remittances of younger relatives for household reproduction, including maize agriculture. For their part, many migrants return to their hometowns to build houses, get married, and live for extended periods, often until their money runs out. This is a process of semi- or temporary proletarianization, in the sense that residents go back and forth between the worlds of unpaid work and paid employment many times over the course of their lives. Residents also undertake multiple income-generating strategies at the same time, which can include both unpaid agriculture and paid agricultural work or off-farm employment. Maize farming by self-identified peasants is a significant activity in San José, but agriculture is also on the decline, and the social relations of production and reproduction are changing. Many households appear to be in transition. For this reason I argue that this strategy of combining corn agriculture and off-farm work is not part of a "new" rurality, at least not in places like the valley, for two reasons. First, agribusiness plants (like poultry hatcheries) and maquiladora factories and workshops have moved into the valley, and residents provide an inexpensive and flexible source of labor for industry. The region is best characterized as an increasingly peri-urban space rather than a renewed rural space. Although country and city have long been interconnected through trade, political projects, capitalist accumulation, and the relations of production (Williams 1973), in recent years capital increasingly blurs the separation between rural and urban spaces. Although I refer to Sanjosepeños as "rural" residents, members of the younger (and largely male) generation spend at least as much time in the urban United States as in their rural Mexican hometowns. These young men, and to some degree young women, are more mobile than their parents or grandparents were.

Second, although labor migration and semi-proletarianization support rural households, many households no longer rely primarily on agriculture. For many residents these changes have happened within one or two generations. In addition to rising costs and lowered corn prices, there have been shortages in irrigation water and a loss of interest and knowledge about maize agriculture among the younger generation. As livelihood strategies diversify, the future of maize agriculture and its campesino producers is increasingly uncertain.

CONCEPTUALIZING GLOBALIZATION

So-called primitive accumulation, therefore, is nothing else than the historical process of divorcing the producer from the means of production . . . And this history, the history of their expropriation, is written in the annals of mankind in letters of blood and fire.
—Marx, *Capital* vol. one, part 8, 874–75

To assert the local is in no sense to deny the *global* character of capitalism (both take place simultaneously, (of course) or to obviate the need to theorize the *abstract* properties (for example, the crisis-proneness) of capitalism. Our (spatial) point is simply how things develop depends in part on *where* they develop, on what has been historically sedimented there, on the social and spatial structures that are already in place there.
—Pred and Watts 1992, 11

The narratives of maize agriculture, and its practices, provide a window onto neoliberal policies and "globalization." Neoliberal globalization has been marked by the growth of transnational corporations, the increased importance of the supranational lending institutions such as the World Bank and the International Monetary Fund, and post-Fordist flexible accumulation strategies (outsourcing, free trade zones, subcontracting, information economies, etc.) that signaled the end of the Bretton Woods accord.[14]

Globalization is not the introduction of new markets and commodities—since from the outset, capitalism has had a "globalizing imperative" and even before the colonial era Mesoamerica had rich market and commodity traditions (Cook 2006)—but rather an acceleration of capitalist accumulation and commodification. Following David Harvey (1989), globalization is an intensification of the "time-space compression" of capital accumulation, whereby the spatial boundaries and impediments to the movement of commodities—which at times includes labor—are increasingly displaced.

While the global institutions and transnational corporations mentioned above suggest that globalization is driven "from above," its processes are also (and could be further) directed "from below." Not only do organized social and political movements shape the direction of globalizing economic processes and state policies, but as we go about our daily lives—as consumers, students, or farmers, for example—acting within the constraints of historically shaped formations and these globalizing processes, we have the potential to affect these processes and policies. Rural producers from the global south, like those of the Tehuacán Valley, are affected by these processes, but they also make decisions about how best to respond to them, they build political alliances, and they make requests and demands for support from regional state officials. At the same time, it should be stressed that social actors are situated differently, and not everyone in the same social location acts in the same way or with the same concrete effects. While agro-food corporations in the Mexican countryside are actors with more political and financial power than campesinos and factory workers of the Tehuacán Valley, that they are powerful actors does not in itself explain the complexity of what is taking place in rural Mexico, nor foretell the future. Agro-food corporations have exerted their influence on the government to avoid paying the over-quota tariffs on corn imports, and they have influenced the design of the recent Biosafety Law, but peasants respond, resist, and react, and Mexican activists and their international networks pressure politicians and state officials to address the problems surrounding the regulation of GM corn imports and field trials.

Earlier critiques of a political economy approach in anthropology argued that the third world was treated as the shore at which the ship of history arrived, in Sherry Ortner's well-known formulation (1984). More recently, Anna Tsing (2000) has argued that in the study of globalization, the global south is frequently treated as the "local" which is acted upon by a process of globalization emanating from the global north. I hope to avoid the pitfalls of such conceptual binaries between the local and the global, tradition and modernity, stagnancy and change, victim and agent, by drawing on insights from the anthropology of culture and political economy, and their overlap with social geography and political ecology perspectives. The work of Pred and Watts cited above, for instance, takes into account both the structure of capital accumulation and its historical, cultural and geographical contingency. In other words, globalization, like any moment in the history of capital, takes form through particular spaces, interacting with and remaking natural environments, social relations, and cultural meanings. This process includes interactions between humans and the natural environment. Transgenes, agricultural pests, and soil fertility, for example, interact with each other and humans in an active and unpredictable manner. This anthropological approach insists upon a historical perspective (the history of particular places and "what is historically sedimented there") and multiple scales of analysis, not simply the local and the global. Ethnographic accounts can help us see how contemporary capitalism works through, and is shaped by, particular places, environments, and cultural practices.

Globalization often conjures up the idea of increased and rapid interconnections, flows, and mobility of people, cultures, capital, and goods, but this imagery can deflect attention from how this mobility is restricted, stratified, or "gated" (Cunningham 2004). While globalization renders national borders more porous to certain movements, such as financial transactions or cultural forms expressed and mediated by the internet, it also entails the regulation of people in motion through heightened border and immigration policies, and has done so particularly since September 11, 2001. This is certainly true of the border between the United States and Mexico, the most heavily traversed border in the world, where on any given day some 4,600 Mexicans are arrested and deported for illegal border crossings (ibid. 338). In the mid-1990s the United States government adopted a new border enforcement strategy and closed off parts of the border, which unintentionally redirected illegal crossers to more dangerous routes. As a consequence, the number of deaths due to dehydration and sunstroke among illegal crossers rose. On average, one such border crosser dies while attempting to make it into the United States for every day of the year (ibid.).

The term "neoliberal" is thus used in this book when referring to globalization to indicate the key role of governments in regulating the movement of people; to signal the role of governments and supranational organizations in establishing the policies, trade agreements, and legislation that contribute to the increased mobility of capital; and to highlight that such policies are contingent and changeable rather than predictable and inevitable. In this regard Ankie Hoogvelt (2001) provides a helpful distinction between the process of globalization and the ideology about the process. The former is "a real historical process which marks, in a sentence, the ascendancy of real-time, trans-border economic activity over clock-time economic activity," or the accelerated space-time compression of flexible accumulation (154). The ideology, which Hoogvelt calls "globalism," reifies the process as being beyond human agency or government responsibility. In addition to suggesting that the profits made through free competitive markets will trickle down to the rest of the population, the ideology portrays the dismantling of market barriers as inevitable. In Hoogvelt's words, there is "the belief in the efficiency of free competitive markets and the belief that this efficiency will maximize benefits for the greatest number of people in the long run" (155). Markets are seen as self-regulating, one of the tenets of eighteenth-century liberalism, but as the critics of neoliberalism have shown, governments play a central role in regulating market deregulation. Governments regulate deregulation through monetary policies and the restructuring of the welfare or developmentalist state (153–55). In Mexico, as we shall see, government proponents of neoliberal policies have argued that the problem facing small maize producers is insufficient access to the market. Free trade was heralded as the process that would bring Mexico into the First World. The assumption continues to be that rural poverty is caused by being left behind during market expansion or left out of economic growth, rather than by uneven incorporation into the ambiguous process of capitalist accumulation and development projects.

Another aspect of globalization worth mentioning is the increased reach and intensity of commodification and commodity exchanges. As pointed out in an anthropological collection on the subject, the study of commodities, or goods produced for exchange, "offers a window onto large-scale processes that are profoundly transforming our era" (Stone, Haugerud, and Little 2000, 1). Commodities involve two orders of value, that of exchange and that of use.[15] While use value pertains to the distinct properties of a good and how those qualities satisfy human needs or wants, exchange measures the value of a thing in relation to other things (Marx 1977, 128). As goods become exchanged in a systematic way, people see them as having a relative value: they measure the worth of one good as equivalent to a certain quantity of another. A key factor in determining value is what Marx called "socially necessary labor time," or the amount of labor (and all the activities and services to train or support that labor) needed under certain conditions to make that good. In short, commodities represent congealed labor time.

The anthropologist Scott Cook points to a three-sided process (exchange value, use value, and symbolic value) through which people determine the value of goods. He argues that pre-capitalist Mesoamerica was made up of "commodity cultures" and had a degree of producers' alienation from the market goods they made and a process of social differentiation. But unlike in capitalist economies, the value of exchange in pre-capitalist Mesoamerica was likely subordinate to symbolic value (2006, 189–90).

Although production is dominated by exchange value in capitalist societies, there are goods and services that are not mediated by the market. Feminists rightly point out that capitalist societies rely on areas of production and social reproduction which are not commodified, such as the everyday domestic work of the household generally performed by women or the subsistence agriculture of peasants (Pearson, Whitehead, and Young 1981). And commodification is not a unidirectional process: there are instances of decommodification when an economic good or service is no longer produced for exchange or by waged labor. Some rural communities in Mexico, for instance, have returned to subsistence agriculture after years of relying on the sale of goods or labor—a process of re-peasantization (Walsh and Ferry 2003).

Under capitalism, as Karl Marx famously argued, labor power itself became a commodity—although the treatment of humans as property, or slavery, did exist before capitalism, and simultaneously with capitalism. The commodification of labor was a consequence of the historical process of separating peasants (and later others) from determining the value of what they produced and exchanged, and separating them as well from the ability to maintain themselves without selling their own labor power. Marx called this process of alienation "primitive accumulation." Since this separation or alienation is ongoing today, David Harvey (2003) prefers the term "accumulation by dispossession." Profits are amassed by stripping people of their access to resources or through the "enclosure" of a communal or open-access resource, undermining people's ability to maintain themselves and their households without selling their labor power. Indeed, scholars point to the current wave of rural displacement as "the great global enclosure of our times" (Araghi 2009). In the Mexican countryside neoliberal globalization and the effects of economic crisis contribute to accumulation by dispossession: the advance of capital in the search of profit separates rural people even further from their means of production. In the southern Tehuacán Valley this is a gendered process, in which young, indigenous women and men are drawn into maquiladora employment in particular ways and young men seek their fortunes across the border, where they are seen as homogeneously "Mexican."

Although rural labor migration helps to raise consumption levels and living standards for some, accumulation by dispossession is an unmistakably violent process. So while studying commodities provides a useful lens onto the process of globalization, partly because commodities "carry cultural messages" (as they do in noncapitalist systems) and "are embedded in political and social systems which they both reflect and help to shape," commodities also signal and embody relations of power and exploitation under global capitalism. They draw our attention to questions of power in terms of both the structural or systematic features of the global economy *and* questions of political agency and the particularities of place.

This book draws on several insights from the anthropological study of commodities,[16] political economy, and social geography—more specifically, the insight that maize can be "diverted" into the commodity phase; that peasants are undergoing "dispossession" as part of the process of capitalist accumulation; and that globalization, as a process of expanded commodification and accumulation, has both the structural features associated with global capital and various place-based contingencies—historical, cultural, political, and environmental.

In the Tehuacán Valley the interactions between these structural features and contingencies are deeply gendered as well as linked to relations of racialized ethnicity. The movement of three commodities in particular is tied to growing economic disparities and the remaking of social and natural environments: labor, water, and maize. Spring water has had a profound influence on the organization of community and the relationships between households of different means. After the Mexican revolution there was renewed conflict over the control and ownership of irrigation water in the

Tehuacán Valley. Some residents were able to amass large shares of the precious liquid. Understanding this history of water is central to understanding the local and regional influences on Sanjosepeños's more recent responses to economic and environmental crisis.

In more recent years globalization has led to a proliferation of commodities flowing in and out of the valley, such as migrant labor, blue jeans made in valley maquiladoras, and various types of maize for food and animal feed. Residents are dependent on both valley and United States employment. As we shall see, access to migrant remittances has enabled some residents to invest in agriculture and machinery while others are left without the funds to do so. Remittances and labor migration have also contributed to higher consumption expectations and the monetization of some agriculture. Maize is a commodity in the valley, grown for the purpose of exchange or sale. However, maize is not always cultivated for the *purposes* of exchange. Households may grow maize with the intention of consuming it in the form of tortillas, but along the way it is "diverted" into the commodity phase, into being exchanged between neighbors or sold for money (Appadurai 1986, 26). Maize is such an important crop in part because it is the cornerstone of the diet and as such can be diverted for exchange in times of crisis or income shortfalls.

This book attempts to balance an analysis of the GM corn debates with an examination of what is taking place in the countryside. To understand what is at stake in the controversy over transgenic corn requires engaging with a topic and a group of social actors (and an anthropological literature on peasants) which for some may seem anachronistic. I argue, however, that in Mexico, the debate about GM corn is fundamentally linked to the future of campesinos and the countryside.

Part I DEBATES

When news of the detected transgenes in cornfields broke, one of the most active groups in the Mexican anti-GM campaign, Greenpeace Mexico, declared corn an issue of "national security" (Boletín 0174, 17 September 2001). Because Mexico is the crop's center of origin, this contamination was also considered the first of its kind to occur anywhere. For activists and critical scientists, particularly those active in the coalition In Defense of Maize, the case of transgenic corn has come to represent much more than the environmental risks associated with gene flow between GM and traditional crop varieties. It represents the damage of unfair trading practices on food sovereignty and small-scale agriculturalists in the global south. The coalition argues that Mexican food sovereignty and quality are being eroded under neoliberal globalization, and employs older symbolic associations between maize and the nation to suggest that Mexican culture itself is at risk. Indeed, while other transgenic crops such as soybeans and cotton have been approved and grown commercially in Mexico, they have garnered little of the scientific, activist, or media attention of GM maize. Corn is a unique crop and food in Mexico because of its cultural, economic, and symbolic importance. We begin by examining how critics challenge the prevalent view of scientific risk expertise in the GM corn debates, before moving on in chapter 2 to explore how the coalition emphasizes food quality and sovereignty. In later chapters we focus on how maize producers themselves discuss and practice corn agriculture in the southern Tehuacán Valley.

chapter 1

TRANSGENIC MAIZE
AND ITS EXPERTS

Maize has united Mexicans.—*In Defense of Maize*,
Manifesto pamphlet, October 2004

Outside of Mexico, the government proudly refers to maize as part
of our culture, but here at home their policies don't support native
varieties.—Interview with biologist at CONABIO,
the Biodiversity Commission, May 2005

When transgenes were found in local maize varieties growing in
the highlands of Oaxaca in 2001, the gap in the regulation of GM corn
became the topic of debate—a gap that Mexican environmental
groups had been pointing to several years prior. While one arm of the
Mexican state placed a de facto moratorium on scientific field trials
of GM corn in 1998, another arm was approving shipments of imported corn from the United States, which included transgenic varieties not labeled or identified as such. Regulators from the Ministry
of Agriculture's Directorate of Plant Health decided that in Mexico,
the center of origin, domestication, and biological diversity of maize,
trials of GM corn were potentially more harmful than beneficial.

Early on in the GM corn debates, some scientists downplayed
concerns about gene flow between landraces and GM corn varieties
by arguing that they were "more related to cultural factors rather
than biological ones."[1] Yet as the corn debates intensified, cultural
concerns began to take center stage. Anti-GM corn activists and
critical scientists affiliated with the emergent coalition In Defense of

Maize emphasized that maize is quintessentially Mexican, the cornerstone of ancient Aztec and Mayan cosmologies, that it was originally domesticated in the area, and that it is now the most cultivated crop and the mainstay of most diets, rural and urban. As one coalition slogan put it, "Without corn, there is no country" (Sin maíz, no hay país).[2]

By employing maize as a symbol of Mexican culture, the pro-maize and anti-GM coalition challenges the official perspective which focuses on the risks of gene flow between transgenic corn and criollos and privileges scientific expertise in evaluating such risks. In the process notions of expertise and risk are contested. The coalition maintains that campesinos who grow maize are affected by regulatory frameworks and agricultural policy and should therefore have a place at the table when decisions are made, but it also contends that campesinos are corn experts. In making this claim the coalition rightly challenges the assumption that smallholder corn agriculture is inefficient because of its low yields and its use of "traditional" technology, such as non–breeder improved seed. As we shall see in chapter 2, this argument refocuses attention on questions of food sovereignty and quality rather than on a market-based logic of efficiency.

Debates about GM foods and crops are dominated by questions and calculations of risk in Mexico and internationally. In his well-know book Ulrich Beck defines risk as "a systematic way of dealing with hazards and insecurities induced and introduced by modernization itself." (1992 [1986], 20). Modern societies are "risk societies" because they distribute the hazards of industrial production and technology, rather than simply the goods and benefits of such production. Since the development of new technologies like agricultural biotechnology can create unintended or incalculable consequences, and in the process outpace existing regulatory frameworks, these technologies are often open to controversy (Herrick 2005). Moreover, contemporary societies are "risk societies" in the sense that the discourse of risk has become a very powerful way to discuss and manage hazards. In the GM corn debates critics of transgenic maize have learned to translate their concerns into the discourse of risk even as they challenge the narrow governmental focus on the risks of gene flow.

While debates about the regulation of transgenic corn are highly polarized, some supporters and participants in the coalition In Defense of Maize—who express their opinions at conferences, in the press, and in anti-GMO campaigns—articulate perspectives about agricultural biotechnology in general that go beyond simple pro or con formulations. The most common example of this tendency was seen among scientists (and to a lesser extent the environmental activists) I interviewed, who felt that transgenic corn should not be imported or commercially released in Mexico, but also mentioned that biotechnology has a place in the improvement of agricultural crops. In other words, these interviewees did not reject all forms of agricultural biotechnology or transgenic crops, but focused on the particular problems and risks associated with maize in the Mexican context. These interviewees also stressed the unique importance of maize in Mexico—which is quite different from its importance to Mexico's NAFTA partners.

In many ways rural Mexico is distinct from an industrial "risk society," but the finding of transgenic corn among native cornfields illustrates that rural areas are nevertheless affected by the "hazards and insecurities induced and introduced by modernization itself" (Beck 1992, 92). In a different body of scholarship, peasant societies have long been discussed in relation to risk, in that peasants devise strategies to contend with the uncertainties of small-scale agriculture and the possibility of food shortages (Scott 1976). In Mexico recent policies and economic conditions have exacerbated the insecurity faced by small-scale agriculturalists, and maize plays a role in peasants' strategies to reduce that insecurity.

To date only a few studies have been conducted on the rural public's perception of GMOs in Mexico. They suggest that when the news of the GM corn controversy arrived in rural communities, it was not accompanied by contextual information or by "interlocutors to respond to questions and concerns" (Larson and Chauvet 2004, 4). In the affected communities in Oaxaca, campesinos reacted with mistrust of the government and scientists, and wanted to know what transgenic contamination meant, in practical terms, for their agricultural varieties and practices (ibid. 13, 25). In the Tehuacán Valley, however, where studies also found evidence of transgenes, the GM corn controversy was not discussed or raised as an issue of concern by residents during my research. While in any society information may be unevenly communicated, shared, or received, this is particularly true in societies characterized by high levels of social inequality, such as Mexico. This chapter therefore focuses on the GM corn debates among participants and interlocutors, including representatives from campesino communities. But for the most part the corn debates have involved activists, scientists, industry representatives, government officials, academics, and journalists, both experts and laypeople. These activist, industry, official, and scientific communities are not homogeneous, nor do they have clearly marked boundaries. Some participants in the corn debates wear more than one hat; for instance, as both maize scientist and government ministry employee, as peasant community representative and local government official, as indigenous campesino and activist, or as university researcher and newspaper columnist who works in alliance with the coalition In Defense of Maize.

While there is an identifiable official government position on the importation, testing, and commercial release of transgenic maize, a position that is embodied in policy and frequently articulated by the Ministry of Agriculture (and that is also the official industry position), the state is not a unitary entity, and different arms of the state often work at cross-purposes. Thinking about the state as a set of political offices and as the making or unmaking of certain forms of political rule draws our attention to the struggles and alliances within and between different ministries, offices, and public servants in regulating transgenic corn imports and tests. Additionally, thinking about the state in this way draws our attention to how laws and policies contain contradictory logics, and how law and policy are inconsistently put into practice and do not always achieve anticipated results. Pundits who planned and predicted a drop in Mexican maize production after NAFTA, for instance, were wrong—or at least have been to date.

The GM corn debates embody competing visions of agricultural development and the future of the countryside. Anti-GM activists call our attention to the political nature of the neoliberal corn regime, despite attempts to present its policies and genetically engineered crops and foods as non-issues. Unlike in some other countries, where the benefits of agricultural biotechnology go largely unquestioned in mainstream discourse and media, In Defense of Maize and its allies contest "biohegemony," or what Peter Newell calls the material, institutional, and discursive power which validates and reproduces the interests of corporate agricultural development (2009, 38).

The GM corn debates are also an arena for rival definitions of biotechnology and biodiversity. Some participants highlight how the novelty of genetically engineered plants has unknown consequences; they point to the risks of gene flow between GM varieties and traditional ones. Conversely, other participants downplay the novelty of genetically engineered plants and argue instead that gene flow between GM and traditional varieties contributes to the biological diversity of plants. The question is whether agricultural biotechnology is seen as an aid to biodiversity and its rural guardians, or as a potential threat or hazard.

DEFINITIONS:
BIOTECHNOLOGY AND BIODIVERSITY

Biotechnology refers to a set of recombinant DNA techniques first developed in the 1970s which use organisms, their parts, or their processes to modify or create living organisms with particular traits. Agricultural biotechnology includes genetic engineering and tissue culture techniques. Genetically modified or transgenic plants are the products of such tools; they are plants whose genomes contain inserted DNA material from other plants or species. Although both conventional plant breeding and farming practices produce new gene characteristics in plants, plant breeding and farming differ from genetic engineering in that they work at the level of the whole plant. Genetic engineering has the capacity, at least in theory, to overcome the sexual incompatibility of different species and to identify, isolate, and relocate any gene from one organism to a recipient plant's genome. In practice, these technologies do not always express the incorporated genetic material or the targeted traits in a predictable or precise manner; genes may function differently in new contexts. Further, there is disagreement over the claim that genetic engineering is faster and more precise than scientific plant breeding (Gepts 2002; Levidow and Tait 1995; McAfee 2003b).

Agro-biotechnology has the potential to aid both subsistence and commercial agriculture in diverse environments; but unlike plant breeding in North America, which was initially developed in public institutions, biotechnology "has been driven by the private sector from the start" (Otero, Scott, and Gilbreth 1997, 259; Kenney 1986). The links between corporate and public research in agricultural biotechnology are increasingly blurred by complex institutional and funding arrangements. Rather than condemn all agricultural bio-technology, some researchers argue for supporting public research, particularly on crops of use and benefit to poor farmers of the global south (Stone 2002).

"Modern" and "Traditional" Varieties

The development of new corn varieties through scientific plant breeding and biotechnology relies on the genetic information from "traditional" varieties found in Mexico and elsewhere. Without tra-ditional varieties conventional agriculture would be unable to adapt to new conditions or pests. The southern corn leaf blight that de-stroyed 15 percent of the corn crop in the United States in 1971 is perhaps the most famous example of the vulnerability of indus-trialized mono-crop agriculture. Cornfields were sown with vari-eties that shared a genetic component susceptible to the new blight. "The epidemics of the early 1970s served to underline a simple but humbling point: although the 'North' (meaning most northern in-dustrialized countries) is grain-rich, it is gene-poor" (Fowler and Mooney 1990, xi). In this way conventional agriculture is dependent upon seed stored ex situ, in germplasm banks and in situ, main-tained and developed in peasants' fields.

Modern or improved corn varieties (also called cultivars) are those varieties developed through scientific plant breeding and biotech-nology. In contrast, traditional varieties—called "landraces" in En-glish and maíz nativo or criollos in Spanish—are varieties maintained, selected, and improved by farmers in the field; they are "crop popu-lations that have become adapted to farmers' conditions through natural and artificial selection" (Aguirre Gómez, Bellon, and Smale 1998, 7). In Spanish, criollo is the commonly used term to discuss traditional varieties. It refers to both landraces and creolized vari-eties, the latter of which are the outcome of an intentional or acci-dental mix of landraces with scientifically improved varieties. The terms "modern" and "improved" are problematic because they im-ply that criollos are not the result of human intervention or improve-ment. This is not so. The development and maintenance of criollos involve a kind of on-farm or in-field plant breeding, albeit one distinct from the scientific plant breeding tradition which emerged in sixteenth-century Europe.

Maize is a naturally hybridizing plant, but a maize "hybrid" is a modern variety that results from the crossing of two different varieties, each of which has first been inbred to the point of being genetically uniform. The first generation of a hybrid variety (called F-1 by plant breeders) has an increased yield, or hybrid vigor. Unlike open-pollinated varieties developed by scientific plant breeders, though, the second generation of hybrid corn (F-2)—that is, the generation that appears after the seed is saved and replanted—exhibits a considerable reduction in yield. Farmers interested in maintaining good yields must purchase hybrid seed for each planting. These varieties also tend to depend on commercial inputs such as synthetic fertilizers and pesticides. Today all commercial transgenic corn varieties are also hybrids.

All seed is both a means of production (as seed to be planted) and a product (the seed to be sold, exchanged, or consumed); but hybridization transforms self-regenerative seed into a non-renewable product. As the sociologist Jack Kloppenburg has explained, hybridization "uncouples seed as 'seed' from seed as 'grain' and thereby facilitates the transformation of seed from a use-value to an exchange-value" (1988, 93).

Hybrid corn played a key role in the establishment and success of seed companies in the United States in the early twentieth century. Today the top ten multinational seed corporations account for over half of the world's commercial seed sales, worth nearly $12.6 billion in revenues (ETC Group 2007). The value of biotech seed and related technology fees is $7.5 billion globally (International Service for the Acquisition of Agri-Biotech Applications 2008). In recent years, large seed companies have come to dominate the sale of seed in Mexico. From the 1960s the state seed producer and distributor, PRONASE, was made responsible for producing and distributing commercial varieties developed by Mexico's public research office on forests, agriculture, and livestock (INIFAP). This arrangement amounted to a monopoly on the seed industry that limited the role of private companies in seed production and sale. But as part of the neoliberal corn regime of the 1980s and 1990s, a seed law was introduced (1991) and PRONASE was dismantled, opening the way for several national and multinational corporations to dominate the market. In 2007 a new seed law took effect which required any seed exchanged or sold—even local criollos—to be registered in the national seed catalogue, or else the person exchanging or selling it will face penalties. This law does not reflect the realities of small-scale, informal seed exchange and sale in Mexico but rather creates a means for large companies to take action against seed producers who challenge their share of the market.[3] Like other aspects of the neoliberal corn regime, this law expands the state's powers in the countryside while simultaneously opening the door for the further accumulation of capital.

With the advance of biotechnology, plant biodiversity is valued as a source of genes for the development of new technologies, crop varieties, and pharmaceutical products. Agro-biotechnology extends the commodification of seed because much of GM seed is accompanied by intellectual property rights (IPRs), requiring users to pay a licensing fee in addition to the initial seed purchase. This fee runs counter to the widespread practice of peasants and farmers to select, save, and even exchange seed for replanting. By charging a fee to use seed saved by farmers, intellectual property provides another way to overcome the free reproduction of seed,[4] or seed's "biological barrier to commodification" (Kloppenburg 1988). The commercialization of seed, including IPRs, is one way that public resources or "the commons" are undergoing privatization or enclosure; it contributes to "accumulation by dispossession," or the accumulation of capital by undermining a group's access and control over the resources that it needs to maintain its livelihood (Harvey 2003, 147–48).

Most attempts to enforce intellectual property restrictions on seed have thus far taken place in the global north. A notable and well-publicized case of patent enforcement grew out of Monsanto's claim that a Canadian farmer, Percy Schmeiser, was growing Roundup Ready canola without a license.[5] In Mexico patent protection is not granted to biological processes, plants, animals, and humans, whereas microorganisms, proteins, genes, cellular lines, antibodies, and pharmaceutical products and microbiological processes can now be patented (Pechlaner and Otero 2008, 363).

Biodiversity

Biodiversity became an international watchword for concerned scientists and environmentalist organizations in the 1980s. It emphasizes the "value" of the natural environment—be it a scientific, economic, ecological, spiritual, or intrinsic value (Takacs 1996). The term can refer to different levels of biological variability: the variability of genes, populations, and species as well as the variability of ecosystems of which they are a part. Policymakers, biologists, the biotech and seed industries, indigenous communities, and environmental activists often define the value of biodiversity differently (Brush and Stabinsky eds. 1996). While industry focuses on genetic materials as an economic resource with potential market value, some environmentalists argue that we must try to save all species from extinction, while others favor a more practical approach which focuses on saving those species whose loss would have the most detrimental impact on human existence (Takacs 1996). These conflicting ideas about the value of biodiversity—and the trade in and control over genetic resources—are disputed at international forums and agreements, such as the WTO's Trade Related Aspects of Intellectual Property Rights (TRIPS) and the Convention on Biological Diversity (CBD). The convention recognizes the importance of "customary uses" of biodiversity for indigenous and rural communities but also supports the commodification of nature by viewing natural resources as best protected through systems of private property rights (Müller 2006b). Market mechanisms are promoted as the way to protect and manage plant genetic resources.

In the case of maize, in situ biodiversity refers to a dynamic process in which criollos are maintained and altered through pollination within and between fields, and through seed selection and exchange among corn growers (Serratos, Willcox, and Castillo-González eds. 1996 [Engl. trans. 1997]). Although the storage of landraces and creolized varieties ex situ in seed banks is a crucial means of protecting maize biodiversity, it is often viewed as an insufficient measure on its own. The CBD, for instance, considers in situ conservation the best approach for safeguarding biodiversity (Glowka, Burhenne-Guilmin, and Synge 1994).

Thus the concept of biodiversity can also include human interaction with ecological systems. This is particularly significant for corn. Unlike some other plants or crops, corn is incapable of sustained reproduction without human intervention. Maize cannot disperse its own seed, which is located under the husk. If an ear of corn does fall to the ground and releases its seed, so many competing seedlings develop that none are likely to grow to maturity.

Mexico is the center of maize biological diversity, home to fifty-nine racial complexes, many criollos, and maize's wild relatives, teosinte and tripsacum (Bellon and Berthaud 2006; Matsuoka, Vigouroux, Goodman, Sanchez, Buckler, and Doebley 2002). There has been some debate about the extent to which the biological diversity of criollo maize and its relatives has declined because of gene flow with, or substitution for, scientifically improved varieties—if it has declined at all. There has been a reported decline in the abundance of teosinte (Wilkes 2007).

The adoption of scientifically improved seed has been limited; it accounts for between 20 and 30 percent of all land cultivated with maize in Mexico, with a much higher percentage in the northern states. The rest of the cultivated land is grown with criollos, both creolized varieties (estimated around 25 percent) and landraces (50 percent) (Ortega-Paczka 2003, 142). Some small-scale farmers try out improved seed or criollos from other regions and cross them with local varieties, either intentionally or by accident (Bellon and Berthaud 2006, 6).

Various factors bring about the loss or abandonment of criollos, including the proximity of farmers to commercial markets and infrastructure (Aguirre 1999 in Turrent and Serratos 2004) and the influence of government programs. Some criollos are more likely to be replaced by other varieties, such as very early- or very late-maturing varieties (Ortega-Paczka 1999). The question of whether the diversity or abundance of criollo maize is on the decline has helped to fuel the scandal over GM corn.

In Mexico smallholder producers with fewer resources generally rely more heavily on criollos than medium-sized or large holder farmers who are more likely to purchase scientifically improved varieties and their inputs. Yet as we shall see, peasants have a number of reasons for preferring criollos beyond a simple inability to afford improved varieties: in some cases they prefer the taste of criollos, or their role in strategies for minimizing risk.

BACKGROUND TO THE CONTROVERSY

The debate over GM corn began in the mid-1990s among Mexican scientists and government regulators with the impending commercial release of transgenic corn in the United States. The Mexican General Directorate of Plant Health (DGSV) of the Ministry of Agriculture began to grant permits in 1988 for scientific field trials of GM crops. The directorate was advised by an ad hoc committee consisting of scientists from various disciplines and government agencies, which became the National Agricultural Biosafety Committee (CNBA) in 1992. The first field trials of corn, conducted by research institutes, were permitted in 1993 and followed by industry trials three years later (SENASICA 2005). The main agro-biotech companies that ran field trials were Monsanto, Pioneer, and Asgrow (a seed company purchased by Monsanto in 1997). Over a period of ten years Mexico approved the commercial release of over thirty-one agricultural GMOs for human consumption, including alfalfa, canola, cotton, tomatoes, soybean, potatoes, and maize (Pechlaner and Otero 2008). No other commercially grown or tested GM crop in Mexico has generated the same level of controversy as transgenic maize.

In late 1998 the CNBA imposed a de facto moratorium on transgenic corn trials because the traits most commonly tested were not of any particular benefit to Mexico (Alvarez-Morales 1999, 91). Members of the committee were also concerned about the possibility that transgenic corn might hybridize with or displace criollos and teosinte (Serratos 1996; Serratos, Willcox, and Castillo-González eds. 1996). This CNBA advisory committee is now known as the Specialized Agricultural Subcommittee of the Inter-Ministerial Commission on Biosafety (CIBIOGEM), which coordinates the efforts and norms of the ministries of agriculture, environment, health, public education, interior, and economy in regulating GM field trials, foods, and commercial agriculture. Previously the Ministry of Agriculture had been the only ministry active in approving field trials and the commercial release of agricultural GMOs.

There is little known about the effects of gene flow between transgenic corn and criollos. Gene flow between corn varieties requires the movement of seed or pollen from one population of corn to another, fertilization between the two populations (or hybridization), and then introgression or incorporation of the new gene into the genome of the recipient population (Cleveland and Soleri 2005). Each step of this process, from seed flow to introgression, is referred to as gene flow (Soleri and Cleveland 2006). The effects of introducing transgenic varieties into agricultural systems are context-specific (Gepts 2002; National Research Council of the National Academies 2002), and concerned scientists believe that more information about gene flow is needed. There are numerous variables involved in the diffusion of a transgene or transgenic variety, and its possible level of harm, or absence of harm, for criollo maize, such as the frequency of pollen flow and the relative fitness of the transgene and the recipient variety.[6]

Like those varieties developed by plant breeders, GM varieties can displace or hybridize with landraces and creolized corn. Some scientists suggest that after a few generations of planting, a transgenic variety of maize would probably creolize and coexist with other varieties. An exception to this can occur when a large area of the new variety is planted and swamps local maize populations (Bellon and Berthaud 2006). The researchers Mauricio Bellon and Julien Berthaud argue that the introduction of a few novel transgenes will probably not decrease the overall genetic diversity of criollos. But the interaction of transgenes with maize populations "create situations that have never been considered in the biosafety risk assessment and management protocols that regulate transgenic varieties in industrialized countries" (ibid. 10).

Other potential impacts of transgenic corn include the development of weeds tolerant to herbicides and pests resistant to transgenic plants, or negative impacts on non-target organisms (Serratos 1996, 69; Turrent Fernández 2005).[7] Once introgressed in crop populations, transgenes are very difficult to eliminate for a variety of reasons. For example, if there was no phenotypic expression of the transgene, farmers would not know which seed to select out (Soleri, Cleveland, and Aragón Cuevas 2006).

Corn Imports under NAFTA

With the implementation of NAFTA, there was a notable rise in corn imports. This rise was the result of various factors discussed in chapter 2, including the dismantling of the long-running public program for the purchase and distribution of Mexican-grown maize (CONASUPO).

These imports also reflect what Michael Pollan (2006) has called a "plague of cheap corn" in the United States: an overproduction of the crop which supplies the industrial food chain, with 20 percent of the harvest exported to places like Mexico (53–55). If we are what we eat, North Americans are now people of corn (23). The average person in the United States consumes one ton of corn a year, but only a fraction of this is eaten directly as corn: most of this crop is consumed in the form of meat like beef, chicken, or pork, since corn is what these animals are now fed; it is also consumed in soft drinks

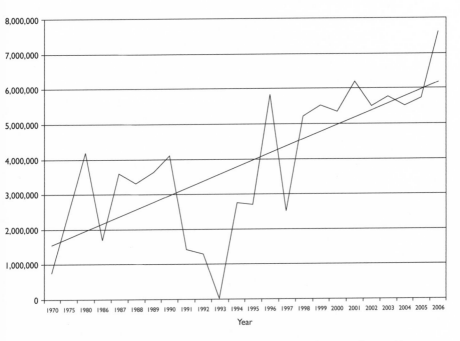

rce: Servicio de Información Agroalimentaría y Pesquera, with additional information from Barkin 2002.

as high-fructose corn syrup and in processed foods (85). The "cornification" of the food system in the United States has had a detrimental impact on the environment, family farms, and consumers' health (118). Farmers for the most part sell their corn for less than the cost to grow it. And although they receive government subsidies, the agricultural system is organized to benefit the large buyers of corn, not the farmers.

The United States is not only the largest producer of maize in the world but by far the largest producer and exporter of GM crops, followed by Argentina and Brazil (ISAAA 2008). The number of countries planting GM varieties went from six in 1996, when such varieties were first commercialized, to eighteen in 2003 and twenty-five in 2008 (ibid.).

In the year 2000 GM corn accounted for 25 percent of all corn grown in the United States. Nine years later GM corn had risen to 85 percent, of which 17 percent was insect-resistant (Bt corn), 22 percent was herbicide-tolerant, and 46 percent consisted of stacked gene varieties which combine both herbicide tolerance and pest resistance (U.S. Department of Agriculture, Economic Research Service 2009). Bt corn expresses the bacterial toxin *Bacillus thuringiensis*, which is poisonous to the European and Southwestern corn borers. These pests burrow into the stem of the corn plant, causing it to fall over. In the United States, once transgenic corn has been harvested there is no mandatory labeling or segregation of it from conventional corn. Moreover, regulation does not consider the risks to areas outside the United States, despite recommendations that it should do so (National Research Council of the National Academies 2002).

Global Governance

Mexican agricultural and food policy since the 1980s has increasingly been influenced by global regulatory bodies, multilateral lending agencies (such as the World Bank and the IMF), trade agreements, and international standards on food safety and biosafety. Previously, Mexican agricultural policy was influenced by international foundations, such as the Rockefeller Foundation in the 1940s. But with the establishment of the World Trade Organization (WTO), its Agreement on Trade Related Aspects of Intellectual Property Rights, and the Agreement on Agriculture in the mid-1990s, Mexican agricultural trade and policy were brought into the sphere of global economic governance (McAfee 2003a, 31). Mexico was one of the first countries in the global south to enact stronger intellectual property legislation (in 1991 and 1997), bringing the country in line with the TRIPS agreement.

The World Trade Organization refers to the Cartagena Protocol on Biosafety (part of the Convention on Biological Diversity)[8] for decisions about GM foods and GMOs and to the Codex Alimentarius Commission (Codex) for food safety standards. The protocol was designed to help the safe transfer and use of living modified organisms (LMOs), or GMOs that can reproduce in the environment. While it does not provide specific risk-management guidelines, the protocol does contain the "precautionary principle," which states that an absence of scientific knowledge about a risk should not hinder actions to reduce that risk (NRC 2002, 64).[9] The protocol thus enables a country to ban the import of GMOs until they are proven safe for human health and the environment and to demand that they be la-

beled as GMOs. A ban may be based on "socioeconomic consider-
ations" and potential risks to biological diversity (Article 26, Secretariat of the Convention on Biodiversity, 2000), although countries
which decide to deny entry to GM imports may be required to provide
the scientific basis for their decision. In Mexico, when the anti-GM
corn coalition first started to mobilize it filed a complaint with several government ministries and offices requesting the halt of GM
corn imports; their complaint was based on the precautionary principle and argued that a failure to halt the imports was a violation of the
Convention of Biodiversity (Grupo de Estudios Ambientales 2007, 62).

In 2003 the Miami Group, made up of the United States, Canada,
Australia, Argentina, Chile, and Uruguay, proposed modifying a provision of the protocol which restricts a country's ability to reject imports
of LMOs intended for food, feed, or processing, for use in contained
places like laboratories, for the production of drugs for humans, and
to pass through a country on their way to another destination.

To preempt the labeling requirements being negotiated by the
parties of the protocol, the NAFTA signatory countries—Canada, the
United States, and Mexico—signed a trilateral agreement in October
2003. Under this agreement[10] Mexico requires that shippers identify cargoes that are known to include GMOs, but a shipment that
consists of less than 5 percent GMOs is considered equivalent to
a non-GMO shipment and does not require identification. Critics
maintain that up to 5 percent of six million metric tons or more of
imported corn annually is an unacceptably large quantity of transgenic corn.[11] In 2004 Mexico also implemented an import ban on
some types of transgenic corn, specifically biopharmaceutical corn
engineered for nonagricultural purposes.

The Corn Controversy

While the moratorium on GM corn field trials was in place, environmentalist groups began to suspect that transgenic corn was making
its way into Mexico in shipments of imported corn from Canada
and the United States. In 1999 Greenpeace Mexico tested samples
taken from ships in the port of Veracruz carrying corn from the
United States and found GM maize among the grain. It launched an
anti-GM corn campaign, as did CECCAM and the Environmental
Studies Group (*Grupo de Estudios Ambientales* or GEA), also based in
Mexico City, and the Canadian-based Action Group on Erosion,

Technology and Concentration (the ETC Group, formerly called RAFI), which established an office in Mexico City in 1999.

In 2001 Ignacio Chapela and David Quist from the University of California, Berkeley, reported that they had found the presence of three distinct transgenic DNA sequences among local maize in the Sierra Norte of Oaxaca: the *Bacillus thuringiensis* (Bt) toxin gene, the cauliflower mosaic virus (CaMV) gene promoter, and the nopaline synthase (NOS) terminator sequence (Chapela and Quist 2001; Quist and Chapela 2002). Gene promoters are DNA sequences that make the transgene function with a plant's genome, telling the gene when and under what conditions to function. The NOS terminator halts or "terminates" the expression of an inserted gene in a plant after the desired protein synthesis has occurred.

Dr. Chapela contacted several Mexican scientists and government offices to share his and Quist's findings before publishing their study in the journal *Nature*, but received a mixed reception. While some colleagues were interested and concerned, others called into question Dr. Chapela's plan to publish his findings (Delborne 2005, 222–25). When the study was eventually published in *Nature* it sparked an international debate about not only the risks of gene flow between such genetically modified corn and criollos but the re-liability of the researchers' study. Although *Nature* had peer-reviewed and accepted the paper, in an unprecedented turn the journal with-drew its editorial support for the study in 2002.

Disagreement over methods is a regular part of scientific inquiry, but Chapela and Quist were unfairly criticized in pro-biotechnology list-serves as "activists"; they were portrayed, along with their meth-ods, as sloppy and biased.[12] The smear campaign was traced to the Bivings Group, which develops internet advocacy campaigns and has Monsanto as a client (Worthy, Strohman, Billings, Del-borne, Duarte-Trattner, Gove, Latham, and Manahan 2005). Al-though some errors in Chapela's and Quist's study did come to light, as the authors themselves discussed in 2002, their finding of transgenic DNA constructs in criollos has not been refuted by scien-tific studies (ibid.).

To verify Chapela's and Quist's findings the Mexican National Ecol-ogy Institute (INE) of the Ministry of the Environment (SEMARNAT) and the Biodiversity Commission (CONABIO) tested samples from different localities in the states of Oaxaca and Puebla and confirmed the presence of the CaMV promoter used in many commercial GM crops (Ezcurra, Ortiz, and Soberón 2002; INE-CONABIO 2002). This study also found transgenic DNA among the grain from some of the government's own rural DICONSA supply stores. As a consequence DICONSA reports restricting its purchase of corn to domestic grain only.

Other studies then followed. Serratos, Gómez-Olivares, Salinas-Arreortua, Buendía-Rodríguez, Islas-Gutiérrez, and de Ita (2007) detected transgenic protein in two maize samples taken from sampled communities in the Soil Conservation Area of Mexico City (the Distrito Federal) using the enzyme-linked immunosorbent assay (ELISA). A condition for planting in the area is that farmers must use non-GM varieties, but since farmers purchase seed that comes into the region, often through informal seed networks, it is hard for them to know whether it is transgenic.

The early studies which found the presence of transgenes were initially ignored and then challenged by the Mexican Ministry of Agriculture and the biotech industry. My concern is not which studies conform to rigorous scientific standards for testing corn for evidence of transgenes, but how particular forms of expertise about maize have been validated or disqualified.

FRAMING THE FINDINGS: IMPROVED NATURE OR POLLUTION?

Proponents of agricultural biotechnology often describe the technology as the latest phase of a long history of human intervention and improvement of crops that began with the invention of agriculture. Downplaying the novelty of biotechnology can undermine any ethical concerns that consumers may have with the use of gene technology to improve crops. In contrast, some GM critics emphasize the novelty of biotechnology, or what they see as the "unnaturalness" of technologies which can transfer a gene from one species to another. Other critics, such as several Mexican biologists and environmentalists whom I interviewed, do not oppose the use of biotechnology in agriculture per se but do oppose industry priorities and weak regulatory measures.

The biotech industry and its proponents also portray agricultural biotechnologies as improved nature. In his study of English agro-biotechnology advertisements, Les Levidow (1991) demonstrates how this technology is represented as improving upon natural evolution in a controlled and predictable manner. Nature without technological intervention is seen as inefficient, messy, and uncontrollable. Technologically sophisticated control over farm inputs is portrayed as more precise and therefore "cleaner" than traditional plant breeding, even though genes often have different functions in different contexts and express traits other than the desired ones when isolated and inserted into recipient genomes (Gepts 2002; McAfee 2003b). The industry also suggests that agricultural biotechnology speeds up the process of improvement when compared to conventional plant breeding, which is not necessarily true (ibid.).

Similarly, in the Mexican corn debates some participants discuss modern corn varieties, including transgenic maize, as a more efficient or improved version of nature. The Ministry of Agriculture and the biotech industry minimized the significance of the studies by Chapela and Quist and by INE-CONABIO by suggesting that the gene flow between GM varieties and criollos is not only part of a natural process of hybridization but a beneficial one.[13] In situ corn biodiversity always depends upon exchange between varieties, and since the 1940s this process of mixing has involved exchanges between native maize and scientifically improved varieties. Critics concede as much, while arguing that the finding of transgenic corn in Mexico is not benign hybridization but rather a form of genetic pollution or contamination (Cleveland, Soleri, and Aragón 2003; Soleri, Cleveland, Aragón, Fuentes, Ríos, and Sweeney 2005).

The minister of agriculture at the time, Javier Usabiaga, initially denied the reports of transgenic corn found in Mexico.[14] He also claimed that genetic flow between GM and landraces is safe and may even be advantageous to landraces by boosting their resistance to insects and herbicides. Similarly, a molecular biologist and director of the public Center for Research and Advanced Studies at the National Polytechnic Institute (CINVESTAV) regarded the finding of transgenic maize among the cornfields of indigenous peasants as part of the natural exchange of genes which improves native varieties. The director's comments were posted and circulated by AgBioWorld, the organization dedicated to promoting agricultural biotechnology: "There is no scientific basis for believing that out-crossing from biotech crops could endanger maize biodiversity. Gene flow between commercial and native varieties is a natural process that has been occurring for many decades. Nor is there reason to believe that these genes will become fixed into landraces unless farmers select them for their increased productivity. In the end, that would result in improving the native varieties" (Dr. Herrera Estrella, posted 19 December 2001).[15]

In contrast to Dr. Herrera Estrella's perspective, one of the maize biologists I interviewed argues that the benefits of gene flow have been misrepresented. Dr. José Antonio Serratos played a role in the early regulation of GM corn field trials. At the time most GM corn was Bt corn, a variety which has little applicability in Mexico: "The debates about Bt corn in Mexico are generally based on studies in the U.S. or Europe. We need to study the question in Mexico. The impact of Bt in the Mexican context and environment is very different. Promoters of biotech say how wonderful it is that Bt corn was found in Oaxaca because it's going to help peasants. But this is incorrect because in Mexico we don't have the pests that Bt is designed to attack. Bt corn is not effective against the types of pests we have in Mexico, but yet Bt corn does have other risks" (Dr. José Antonio Serratos, 28 January 2002). The maize scientists I interviewed who were critical of transgenic maize, like Dr. Serratos, pointed to a dearth of information and scientific study on gene flow between GM corn and native varieties and between maize and its wild relatives; inadequate government support and funding for this sort of study and for in situ conservation in the Mexican countryside, where small farmers cultivate corn; and the unsuitability of existing transgenic varieties to address problems specific to Mexico. While Dr. Serratos was critical of the regulatory gap in Mexico and utility of GM corn field trials for Mexico, he did mention that he believed biotechnology has a role in improving agriculture.

The industry and government perspective that GM corn can be beneficial to Mexican native varieties also tends to view improved varieties, including GM seed, as a remedy to the inefficiency of "traditional" maize production. The discovery of transgenic maize in Oaxaca and Puebla is seen as part of natural hybridization, interaction, and exchange; but at the same time, the problem of small-scale peasant agriculture and maize varieties is that they have not interacted enough with improved seeds and other "modern" technologies. Traditional agriculture is deemed inefficient because it is low-yielding, labor-intensive, and largely based on traditional rather than modern technologies, such as tractors and breeder- or biotech-improved seed. Modern varieties generally produce higher yields, particularly in irrigated areas. Mexican corn yields average 2 tons per hectare in rain-fed areas and 5.8 tons per hectare in irrigated areas, with yields in the north closer to those in the United States (where yields average 8.5 metric tons per hectare and can reach 12 tons; Nadal 2000a; Zahniser and Coyle 2004, 5).

Agro-biotechnology is also portrayed by the industry as the key to averting food shortages and hunger caused by population growth, particularly in the global south. The claim is that biotechnology can increase productivity to match growing demand. But this argument incorrectly posits hunger as the result of population growth and food shortages rather than a combination of factors, including distribution and access to resources. It also assumes that technology is scale-neutral: that it is equally as useful to smallholder subsistence farmers as it is to large-scale commercial farmers. Critics have demonstrated that seeds improved by scientific plant breeding or biotechnology are not in fact scale-neutral, since they tend to require costly inputs. Although improved varieties produce high yields, they depend on expensive irrigation, synthetic pesticides, tractors, etc. This criticism originally emerged in Mexico and elsewhere after the Green Revolution of the 1960s (Barkin and Suárez 1983; Hewitt de Alcántara 1976; Shiva 1991). High-yielding varieties are more "efficient" when the other costs—monetary, environmental, and social—are ignored.

Mexico is at times said to be self-sufficient in white corn—the type of corn preferred for food—but this is misleading, because the growing corn flour industry which manufactures the ready-made tortilla mix and packaged tortillas eaten by many Mexicans relies on yellow, often imported, corn. Tortillas made from corn flour dominate the market in urban Mexico. Yet national self-sufficiency in maize is possible with the right policies and scale-appropriate technology (INIFAP cited in Appendini 1994, 153; Turrent, Aveldaño Salazar, and Moreno Dahme 1996).

Contaminated Corn and Tainted Tortillas?

Activist narratives about contamination and pollution have at times included food scare tactics. The title of a press release by the ETC group, for instance, "Contaminated Corn and Tainted Tortillas: Genetic Pollution in Mexico's Centre of Maize Diversity" (23 January

2002), implies that consumers could get sick from eating GM tor-tillas, which is certainly not so in the short or medium term. Since Bt varieties have been consumed since the mid-1990s and Bt spray has been used as a popular pesticide on plants over the years, the regula-tory community considers it safe for human consumption. Even so, while there is no evidence that the most common transgenic corn varieties are detrimental to human health, the Mexican Academy of Sciences contends that there is no "absence of risk."[16] The task of determining the long-term effects of eating transgenic maize is a difficult one, particularly considering that our diets are complex and that GM corn is consumed as an ingredient in many other foods such as industrial-made tortillas in Mexico[17] and soft drinks made with high-fructose corn syrup in the United States and Canada. Anti-GM activists call for labeling GM foods as a way of achieving more transparency in the regulation of GM crops and foods, and so that consumers can decide whether they want to support or eat GM ingredients.

In Mexico GMOs are officially evaluated case by case. In practice the Ministry of Health follows the analysis and decisions made in the United States on the safety of GM crops and foods for human consumption. Unlike pesticides and food additives, which are tested on laboratory animals in the United States, foods are evaluated in both countries based on the model of substantial equivalence. The assessment of whether a food is safe involves an evaluation of the food's proteins, fats, carbohydrates, vitamins, minerals, toxins, and allergens. If a GM crop or food made from it is found to be "sub-stantially equivalent" to a non-GM food that is safe, then the GM food is not subject to further testing or more stringent regulation (McGarity 2007, 131). Under the new Biosafety Law (2005) in Mex-ico, discussed further below, the approach to labeling GMOs for human consumption is one which combines substantial equiva-lence with the precautionary principle, but ultimately favors the former (Pechlaner and Otero 2008, 364).

There are some transgenic varieties, like Starlink corn, which may cause allergic reactions in humans. In 2000 corn taco shells made with Starlink corn (Bt Cry9C) were recalled in the United States. Approved as animal feed but not for human consumption, the variety made its way into the food system in the United States and beyond.[18] Countries like Mexico, which have fewer resources available for regulation and monitoring, are at least as vulnerable to such mistakes. Tests undertaken by a group of Mexican NGOs found evidence of the banned Starlink variety in Mexican cornfields (CECCAM, CENAMI, ETC Group, CASIFOP, UNOSJO, and AJAGI 2003).

Anthropologists point out that foods which blur or mix distinct cultural categories are often seen as impure or dangerous (Douglas 1966). This sense of danger is particularly heightened with GMOs because they are indistinguishable from non-transgenic varieties in the field or GM-free foods at the grocery store, yet they are prepared by using technologies which can transfer genetic material from one species to another. In their analysis of media representations of recent GM food scares, particularly in Europe, Fitzgerald and Campbell (2001) argue that the portrayals of GM food by both the industry and its critics are nostalgic responses: industry expresses a nostalgia for the Enlightenment project of progress by focusing on feeding the world through technology, while the critics yearn to return to a romanticized past when food was pure and safe. But in Mexico the concerns about transgenic corn are more than expressions of "cultural anxiety" and "nostalgia" for a simpler past. These concerns must be situated within the social, political, and economic changes brought about by the neoliberal corn regime, which emphasizes privatizing state duties and agencies, and the shift toward maize imports through liberalized trade. In other words the concerns about maize-based foods, agriculture, and biodiversity are related to larger changes in trade and agricultural policy—something that many anti-GM coalition participants have argued and that this book attempts to demonstrate.

DEFINING RISK AND ITS EXPERTS

In Mexico official accounts of the findings of transgenic maize in Oaxaca and Puebla tend to disqualify in several ways the concerns and expertise of maize producers, activists, environmentalists, and scientists critical of GM corn imports. As mentioned above, some experts minimized the concerns of critical scientists and activists about transgenic corn by promoting biotechnology as a more efficient version of nature, or the exchange between GM and traditional varieties as part of a natural and beneficial process of biodiversity. They argued this even though Mexico does not suffer from the pest that the most common type of transgenic corn at the time, Bt corn, is designed to combat.

Secondly, the focus on the risks of gene flow renders other concerns and contexts less relevant. Transgenic crops are generally evaluated in international agreements and regulatory bodies by using a framework of risk and relying on objective, scientific criteria to evaluate risk. Yet as researchers on the social construction of environmental risk have pointed out, this framework occludes or denies the ways risk assessment is value-laden or influenced by cultural norms or priorities (Levidow 2001; Wynne 2001). Risk assessment for GM crops focuses on the likelihood of gene flow between GM crops and non-GM plants, and the harm that may result. But this focus often trivializes non-risk concerns like ethics, food sovereignty, quality, taste, and farmers' livelihoods (Wynne 2001). It also posits scientists as the only experts qualified to determine the potential harm of GM crops, treating lay concerns as "politically real but intellectually unreal" (ibid., 452). In other words, the concerns of the lay public are acknowledged to have very real political consequences but are seen as based on scientific ignorance. In the Mexican corn debates concerns about transgenic corn are frequently portrayed as unscientific or ideological.

In her work on the French anti-GM movement, the anthropologist Chaia Heller (2002) points out that activists challenge the boundary between scientific "experts" and lay "publics." Activists and concerned scientists criticize the discourse of risk as the central frame for evaluating and debating GM crops, or at times they reproduce the frame of risk in arguing that the risks are too great to allow for field trials, commercial release, or the consumption of GMOs (Heller 2004). Regardless of their emphasis, they question what constitutes "expertise" in the evaluation of risk.

In the United States the risk management system used to evaluate GM crops does not view transgenic crops as posing any novel risks; it evaluates GM crops based on an existing model for invasive species. Risk assessment consists of four main steps: identifying a potential hazard; estimating the likelihood of exposure and harm caused by exposure; assessing the harm; and rectifying the problem by reducing exposure to the harm (Cleveland and Soleri 2005; National

Research Council of the National Academies 2002). There are a few problems with the risk management system employed in the United States. The identification of hazards is based on the principle of substantial equivalence, whose definition and application can be subjective and arbitrary (Cleveland and Soleri 2005). If a GM crop is judged substantially equivalent to an existing conventional variety, no further risk assessment is required. Moreover, the risk assessment process does not consider the potential environmental impact of GM crops in other countries, nor evaluate their effects on crop biodiversity (National Research Council of the National Academies 2002; Snow 2005).

The United States government has promoted its risk management system and GM crops as safe and appropriate for countries of the global south, even though many countries in the global south do not have the resources to replicate the regulatory system of the United States (Thies and Devare 2007). There is also an absence of data on the level of invasiveness of GM crops in the global south, where the centers of crop biodiversity are located (Cleveland and Soleri 2005). With regard to GM corn, concerned scientists, academics, and activists argue that the Mexican environmental, political, and cultural context is significantly different from those of the United States, Europe, and Canada and that the differences change the process and standards for risk evaluation.

In addition, small-scale Mexican agriculturalists may evaluate risk in ways distinct from those of scientific experts or the government. In their research in rural Mexico, Cuba, and Guatemala, Daniela Soleri and her colleagues found that 66 percent of the agricultural producers interviewed "felt that [the process of] transgenesis per se was acceptable" (Soleri, Cleveland, Aragón, Fuentes, Ríos, and Sweeney 2005, 148–49). Yet despite their positive attitude toward the process of agricultural biotechnology, when asked their preference without having the varieties identified as "local" or "transgenic," the farmers chose locally available varieties with a low but stable yield over an improved variety that starts with a higher yield and decreases over time, needing to be occasionally replaced by buying new seed.[19] This study demonstrates that small farmers in Mexico (and in Guatemala and Cuba) evaluate the risk of GM crops in ways distinct from the dominant mode of risk assessment, and that they may evaluate the process of agricultural biotechnology differently from the products of such technology.

In another study two Mexican researchers, Michelle Chauvet and Jorge Larson, found that when they conducted workshops in the Sierra Juárez of Oaxaca to communicate the results of scientific studies on transgenic maize, some campesinos felt that they had been the guinea pigs for this new technology.[20] To remedy this Chauvet and Larson recommend that the results of scientific studies be communicated to small producers in an accessible manner and that campesinos have the right to make their own decisions about whether to use the technology in question, rather than have the decision made for them, in this case through inadvertent introduction. In other words, their work contests the notion that expert knowledge of scientific risk analysis is the only knowledge or perspective that should be taken into consideration in regulating transgenic corn in Mexico. It also raises the question of how best to communicate the results of scientific studies to the lay public.

A third way that the concerns of In Defense of Maize were discredited was by calling into question the scientific methods used to determine the presence of transgenes. A restudy of the Oaxaca sierra which did not find detectable transgenes in local landraces of maize (Ortiz-García, Ezcurra, Schoel, Acevedo, Soberón, and Snow 2005) was used differently by the competing sides of the GM corn debates. While some found serious sampling problems (Cleveland, Soleri, Aragón Cuevas, Crossa, and Gepts 2005), others used the study to bolster criticisms of earlier studies, particularly that of Chapela and Quist. Industry representatives heralded the restudy as proof that fears about maize biodiversity had been blown out of proportion. In an interview Dr. Solleiro of UNAM, former director of AgroBIO, said that based on the study, "It is proven that the transgenes do not remain in the environment" (5 July 2006, UNAM campus, Mexico City). The director of AgBioWorld went so far as to condescendingly title his written reaction thus: "Duh . . . No GM Genes in Mexican Corn" (Prakash 2005). Yet in my interview with one of the study's authors, Sol Ortiz García, she was dismayed about how both the environmental NGOs and industry were interpreting the study:

> Maybe it was naive for us to assume that the NGOs would be happy with our results that we didn't find transgenic corn in Oaxaca. It isn't that there aren't transgenes, we never said there weren't, but we said that at the levels we considered for detection, we didn't detect any

transgenes. For this reason, we called the article "Absence of *detectable* transgenes in local landraces." Almost immediately people arrived at the conclusion that this meant there was no GM corn in the area, but what it really meant is that the transgenes had lowered their frequency. We thought that this was to be taken favorably by various NGOs. But curiously, it was the opposite. The NGOs questioned our article and unfortunately, I think, many did so without understanding what we were saying. We never said that the transgenes had disappeared. And on the other hand, we felt that industry used the article for their own purposes, "See, there are no transgenes, nothing happened." . . . In the article we explicitly state: these findings apply to this region for the years that we took samples. That is to say, these are the results of right now. (11 July 2006, INE office, Mexico City)

When nonscientists have used commercially available kits to test for proteins expressed in transgenic corn, their results have been discounted. In 2003 a coalition of Mexican NGOs that had worked with various rural communities to collect corn plants and seed for their own testing presented their findings. They found that GM corn was not confined to the states of Oaxaca and Puebla, but was much more widespread (CECCAM, CENAMI, ETC Group, CASIFOP, UNOSJO, and AJAGI 2003). In some areas they even found evidence of Starlink corn made by Aventis, which had been recalled from sale for human consumption in the United States. Their tests were discredited among industry and scientists for three main reasons: their test samples were not shared, other test methods were not used,[21] and the tests were not conducted by scientists. Government scientists had asked the NGOs for the tested samples and the locations from which they had been taken, so that they could return to the areas and verify the findings with more reliable testing. But they were not given the locations, or any samples. When I raised this issue with Ana de Ita from CECCAM, one of the NGOs that had taken part in the campaign, she explained that the rural communities did not want to be identified: "The reason we didn't release *where* we found evidence of GM corn, was because the communities themselves didn't want this to be known. The majority of them didn't. It wasn't like in Oaxaca where the communities were public and politically organized about transgenic corn. Also we didn't hand over the 'seeds' for retesting because we didn't test the seeds, we tested corn plants. . . . I'm not a biologist. But we did have help from scientists in figuring out how to conduct the tests. We used the protein detector kit" (24 July 2006, CECCAM office, Mexico City).

Their tests were also discounted because those who conducted them were nonscientists seen as lacking in expertise. In another interview, the biologist Dr. Antonio Serratos mentions this and explains his own perspective, in contrast to that of pro-GM corn scientists:

> The NGOs used kits to detect protein in plants and found positive results [in 2003]. For me, this study to detect protein was reliable and the measures were adequate for a preliminary evaluation. But we need a more profound study. However, the Mexican scientific community has largely rejected these studies because they are not conducted by scientists. This includes CINVESTAV [the Center for Research and Advanced Studies of the National Polytechnic Institute]. It's not that these scientists are insulting. They don't trust the results of NGOs for various reasons; "It's not in a lab, it's without *técnicos*, without controls." But if the NGOs followed the protocol, then we cannot reject these results—whoever conducted the tests. We cannot negate the ability of campesinos to use these kits. The scientific community has rejected the results because they believe the people do not have the capacity to conduct the tests—the scientific rigor and so on. We cannot exclude results because X or Y person did the test. I would not take their results as *certification* that there are transgenes [*transgénicos*] present, but they are useful as preliminary findings." (30 April 2005, Mexico City).

Dr. José Sarukhán of the Ecology Institute of the National Autonomous University of Mexico (UNAM) responded to my question about the studies by saying that the real difference in opinion was not within the scientific community but within the government:

> "[Question: Were the NGO studies accepted or rejected by the Mexican scientific community?] I don't think that we have a big problem in the scientific community here . . . There could be differences in opinion [among scientists] regarding whether a method is more appropriate or to say that there are some results that are more reliable than others, but these differences of opinion are normal in such cases. That was not a problem here [in the scientific community]. The problem lies in the public sector, with various government offices, particularly the Ministries of Agriculture and of the Economy. They have taken the position—

a position I believe is incorrect—which blindly supports industry. I am not suggesting that they turn against the biotechnology industry, but that they should take into account the social aspect of GM corn, not only the social aspect relating to the campesinos who live from land-race corn . . . but also how this relates to consumers. (4 July 2006, UNAM campus, Mexico City)

Dr. Sarukhán explains that differences in opinion over results and methods are normal occurrences in science, but that the criticism from the ministries of agriculture and the economy in this case is based on their uncritical support for the agro-biotech industry. While many participants in the corn debates agree that transgenic maize is but one issue of many that need to be addressed, the government position delineates what it sees as the appropriate focus—risks due to gene flow—and expertise in the debates. The official position, particularly that of the Ministry of Agriculture, has questioned (at the earlier moments of the debate on the presence of transgene constructs in Oaxaca) the evidence that gene flow occurred and the methods used to test for gene flow; in those cases where gene flow has been acknowledged to occur, the official position is that its benefits outweigh any potential harm.

Several scientists from a government agency embroiled in the debates over regulation explained during an interview that the controversy over GM corn has been fueled by misinformation, because nonexperts are dominating the discussion: "The debate should be among agronomists in academic circles, rather than among ecologists and activists or civil society [as it is now]. The debate should be among peasant organizations rather than conservationists. This is a serious absence in the debate . . . The focus on GM is a smoke-screen. It just clouds the issues. We need non-ideological investigations into gene flow" (Mexico City, 11 May 2005). By associating civil society with ideology and agronomists (or science) with neutrality or objectivity, these public servants delegitimize the voices and concerns of nonscientists, with the notable exception of peasants. In Mexico there has been a strong counter-tradition among agronomists, ecologists, and anthropologists which validates peasant agricultural knowledge despite the dominant trend within the Ministry of Agriculture and other government ministries to consider peasant practices and knowledge backward and inefficient. This counter-tradition gained popularity as part of post-revolutionary narratives of the early twentieth century, which celebrated the peasantry (see chapter 2). Additionally, anthropologists outside the Mexican tradition have also sought to validate the knowledge practices of peasants and indigenous groups, along with their folk taxonomies and classifications (Gupta 1998; Nader 1996; Richards 1989). These two traditions, in addition to the current transnational slow food and peasants' rights movements, among other alter-globalization movements, inform the perspective about peasants' expertise in the corn debates today. As the above quotation illustrates, campesinos may be considered maize experts, even among some government officials who call for excluding the rest of civil society from the debates on regulating transgenic corn tests and imports.

As controversy grew over how best to test for transgenic DNA and whether gene flow between GM corn and criollos was a form of pollution or beneficial hybridization, there were also debates about which scientific experts should be represented on biosafety regulatory bodies. One maize scientist told me that the Biosafety Committee, CIBIOGEM, was dominated by pro-biotechnology scientists and defined corn expertise accordingly: "CIBIOGEM is organizing an elite group on the issue of GM corn. They are defining the parameters of the issue, defining who the experts are, the 'real' experts. Look at the composition of the Consultative Committee of CIBIOGEM, at which scientists are on it. A panel designated by CONACYT decides who is on CIBIOGEM, and if you are rejected from the selection process, they won't tell you on what grounds. People who were critical of biotechnology or GM corn were not chosen. Rather, people with no experience on the issue [of maize] were chosen or those who had experience with companies and lobbying" (maize scientist, Mexico City, 2 November 2000).

The inclusion of a researcher who was also an agro-biotechnology industry representative on the Biosafety Committee has been viewed as a conflict of interest by some activists, academics, and plant scientists. Although there is no official industry representative on CIBIOGEM, Dr. José Luis Solleiro was a member of the committee while also the general director of AgroBIO Mexico, which was founded by the biotech corporate giants DuPont, Monsanto, Sygenta, Bayer, and Dow to promote agricultural biotechnology. In an interview Dr. Solleiro suggests that the biotech industry is a part of civil society, and as such its inclusion in regulatory bodies helps promote a diversity of opinion. When I asked Dr. Solleiro how he responds to criticisms that agro-biotech corporations should not have a voice on CIBIOGEM, he responded by saying:

> I'm sure those criticisms are directed at me because I was a member of the Advisory Committee [of CIBIOGEM]. But in various consultative groups Greenpeace was also there. That is, I believe that the fair thing is that in such groups where they debate the issue and make recommendations, that there is representation from all interested parties; that they don't exclude anyone. And if we are talking about money, how much funding does Greenpeace have? Did they tell you? Where does their funding come from? Why don't they say? I said to a representative from Greenpeace when we were on TV together, "Why don't we tell the public how much money we manage and where it comes from? You are always talking about transparency, so why not reveal now, to the public, in front of the cameras, how much money we manage and where it comes from?" [Question: So where does Greenpeace's funding come from?] Greenpeace's anti-GM campaign is financed by producers' organizations and companies from Europe. They receive a lot of funding, a lot more than AgroBIO. We've never been able to advertise on TV like Greenpeace does. . . . AgroBIO only has five employees. (5 July 2006, UNAM campus, Mexico City)

While Dr. Solleiro and some of his colleagues point to the influence and funding behind transnational environmental groups such as Greenpeace, members of In Defense of Maize argue not only that it is a conflict of interest to have transnational corporations represented on state regulatory bodies, but that the voices of these corporations are already heard in the halls of the Mexican senate. Not "all interested parties" have the same power to mobilize resources in Mexico. Private companies have resources above and beyond those of representative organizations like AgroBIO to hire lobbyists and influence public policy. The argument made by representatives of the biotech industry is an interesting one, because it attempts to redefine "civil society" in a way which includes the voices of the world's largest corporations.

The selection of experts for representation on regulatory bodies in Mexico is also contentious because of the country's history of government corruption and political clientelism. There is a deep distrust that government institutions will function in the interests of democracy or transparently. In Defense of Maize reflects this distrust in state institutions when it calls for transparency in state policy and regulatory decision making.

FROM RISK TO CULTURE

One of the turning points in the anti-GM corn campaign was the forum held by In Defense of Maize in Mexico City in January 2002, precisely because it presented an alternative account to the official Ministry of Agriculture's and biotech industry's focus on gene flow. At the forum over three hundred academics, activists, farmers, and NGOs (ANEC, CECCAM, CENAMI, ETC Group, Food First, GEA, Global Exchange, GRAIN, UNORCA, among others) concluded that the problems facing Mexican corn were much larger and more complex than the issue of transgene contamination. So while the controversy over transgene contamination has garnered much attention from the media and activists, it is not, as In Defense of Maize has pointed out, the only challenge faced by maize producers or maize biodiversity in Mexico. At the forum and similar ones that followed, a broader campaign emerged. The forum was a turning point because it presented the results of the study by the Mexican National Ecology Institute (INE) and the Biodiversity Commission (CONABIO), which found evidence of transgenes in criollo corn-fields in Oaxaca and the Tehuacán region of Puebla.

Some of the key organizations and individuals involved in the forum had previous experience working together, particularly around NAFTA and later on bioprospecting agreements.[22] One of the first environmentalist groups in Mexico, the Grupo de Estudios Ambientales (GEA), established in 1977, has been an early and active participant in the In Defense of Maize campaign. The coalition created new networks, and strengthened existing ones, between environmental NGOs like GEA and campesino groups, academics, scientists, and indigenous rights groups, explicitly around the question of maize and all that maize symbolizes: the fate of the Mexican countryside, rural subsidy cuts, increased dependency on food and grain imports, the difficult road to sustainable agriculture, rural outmigration, the rise of the corporate seed industry, and the biological diversity of the most widely cultivated and consumed crop in Mexico.

The anti-GM, pro-maize movement also dovetailed with peasant organizing. While corn production was often an issue in previous movements because of its association with the Mexican countryside, it took center stage as a symbol of rural Mexico in crisis and under threat. In December 2002 a coalition of fourteen peasant groups and 100,000 protesters opposed to NAFTA and neoliberal policy came together under the banner of El Campo No Aguanta Más (the countryside can't take it any more) and organized an impressive series of protests that took over the center of Mexico City. The coalition demanded, among other measures, the renegotiation of NAFTA's agricultural chapter and the immediate halt of GM corn imports. These protests were preceded by years of organizing for land rights, for greater attention to the plight of Mexicans who migrate to the United States in search of work, and against efforts by the IMF to promote economic restructuring, trade liberalization, and cuts to rural subsidies.

The government began negotiations with the El Campo No Aguanta Más movement, promising to renegotiate NAFTA's agricultural chapter or provide some compensation to producers. An Agreement on the Countryside (El Acuerdo Para El Campo) was signed in April 2003, but critics charged that it watered down the movement's concerns about the finding of transgenic corn.[23] Divisions arose between groups in the coalition that focused on immediate goals and others that focused on changing government policy. The government ended negotiations, claiming that it had made progress on most of the agreement.

While political organizing around indigenous rights and peasant livelihoods has a long history in Mexico, in the 1990s these issues were increasingly linked with transnational activist networks working on food sovereignty, peasant and indigenous rights, environmentalism, and alter-globalization. When the GM corn controversy unfolded it became a cause célèbre among international NGOs and networks such as the Via Campesina, Global Exchange, the ETC Group, and Greenpeace.

Members of the pro-maize campaign widened the scope of the debate over gene flow between GM corn and criollos to include other challenges facing the countryside. In doing so they argued that Mexican culture itself was at risk of displacement. While some participants spoke in terms of risk to advance their position, they questioned why in evaluating risk scientific expertise should be privileged (Heller 2004). At pro-maize forums the privileged experts are not solely scientists and regulators, or the activists, academics, and environmentalists who study the issues: they include the campesinos and indigenous farmers who cultivate, process, prepare, and select maize. Much as the anti-GM campaign in France portrayed French farmers as artisans central to the survival of the national cuisine and culture, as shown by the anthropologist Chaia Heller (2002), in the Mexican campaign maize agriculturalists were portrayed as the producers and guardians not just of traditional corn varieties but of national cultural practices and traditions. This notion that peasants are both cultural producers and maize experts is reflected in the writings of In Defense of Maize supporters, such as the following newspaper excerpt by the anthropologist and public intellectual Armando Bartra: "El Popol Vuh and la Suave Patria, emblematic texts of indigenous and mestizo Mexico, emphatically suggest that the Mexican countryside is much more than a great producer of foods and primary materials for industry. Peasants not only cultivate maize, beans, chile, or coffee, they also cultivate clean air, pure water and fertile land; biological, social and cultural diversity; a plurality of landscapes, smells, textures and tastes; a variety of dishes, hairstyles and attire; a great many prayers, sonnets [sones], songs, and dances; peasants cultivate the inexhaustible multiplicity of uses and customs that make us the Mexicans we are" (Bartra, "What Purpose Does Agriculture Serve?," La Jornada, 21 January 2003). Here we see that maize is a symbol of rural and indigenous cultural practices and traditions, as well as Mexicanness more broadly, "the inexhaustible multiplicity of uses and customs that make us the Mexicans we are."

Some indigenous producers who participated in the forums of In Defense of Maize presented the issue of transgenic corn imports as one of indigenous cultural autonomy. For instance, Aldo González, a Zapotec farmer and community representative from the Sierra Juárez, Oaxaca, was quoted in a press release as saying: "The contamination of our traditional maize undermines the fundamental autonomy of our indigenous and farming communities because we are not merely talking about our food supply; maize is a vital part of our cultural heritage. The statements made by some officials that contamination is not serious because it will not spread rapidly, or because it will 'increase our maize biodiversity,' are completely disrespectful and cynical."[24] Maize is emphasized here as a "vital part" of indigenous cultural heritage. Similarly, other peasant participants (who may or may not be indigenous) have portrayed maize as the cornerstone of their culture, and themselves as maize experts. At a workshop organized by In Defense of Maize and NGOs to help farmers from Guerrero, Veracruz, Oaxaca, Chiapas, Michoacán, Jalisco, Chihuahua, and Puebla devise community strategies to protect criollo maize, participating farmers concluded the event with the following statement: "Maize is the fundamental pillar of our resistance and we defend it with the strength of women, men, the elderly and children, because we are of maize. Alongside us, the diversity of maize has grown, enriching the agriculture of Mexico and the world, that only we can know how to heal and decontaminate. The Mexican government has demonstrated that they only want to protect the interests of transnational seed companies; for us maize is not a business: it is our life, and as such, we will defend it" (8 December 2005). In this view scientists and regulators are not the privileged experts in evaluating the risks of transgenic maize. Rather, rural and indigenous farmers are maize experts, the creators and guardians of maize-based agro-diversity and cultural traditions and practices.

IN DEFENSE OF MAIZE

Several interviewees likened the campaigns of In Defense of Maize as the work of ants (el trabajo de hormigas): a vast number of participants work on their own projects individually or in groups, and together their activities produce results. Beyond the first forum sponsored by In Defense of Maize, and its follow-up forums, these projects have included disseminating information to the media, lobbying government ministries, and organizing workshops to communicate the results of studies on transgenic corn to peasants and to define and explain what transgenic corn is.[25] For instance, GEA and UNORCA, together with the Autonomous University of Guerrero, organized a forum in Guerrero in September 2004 entitled "In Defense of Our Maize" to provide information to maize farmers about GMOs and develop strategies for promoting peasant agriculture and food sovereignty. The forum was attended by over 350 people and produced a declaration which called for the right to live with dignity in the countryside, the right to a healthy environment, the right to produce and consume safe, healthy, and diverse foods, the right to reproduce and exchange seeds, the right to information and free elections, and the right to participate in government decisions

(Grupo de Estudios Ambientales 2004, 169). Alluding to the well-known slogan of the World Social Forum, "Another world is possible," the signatories included in their document the phrase "A Guerrero without GMOs is possible" (167).

In addition to organizing forums, demonstrations, and workshops aimed at outreach to campesinos, a network of GM-free tortilla makers based in Mexico City was created and Greenpeace Mexico has produced a list of foods free of GMOs. There has also been a museum exhibit on maize which resulted in a book publication and GM-free food fairs.[26] The Oaxacan government reviewed the possibility of prohibiting the entrance of transgenic corn into the state, and in 2005 the state congress declared Oaxaca an in situ germplasm bank for maize.[27] Other local and state governments have decided to become GM-free zones, such as in the state of Tlaxcala and the municipal government of Mexico City.

While organizing at home, Mexican critics of transgenic corn presented their perspectives at international venues. In 2002 critics presented their case at the World Social Forum and the second World Food Summit in Rome. At the summit Mexican producer representatives from UNORCA and the Sierra Juárez, Oaxaca, spoke about how GM corn poses risks to food supplies and the culture of indigenous Mexicans (McAfee 2003a, 29). Perhaps most important to the critics' efforts was the engagement of the Commission on Environmental Cooperation of North America (CEC), established as part of a side agreement to NAFTA to resolve conflicts related to trade and the environment. In 2002 environmental NGOs and Oaxacan indigenous and peasant groups asked the CEC to investigate the effects of imported GM corn on their communities. The CEC responded by requesting scientific background papers and setting up a series of consultations. After the consultations and a public symposium in Oaxaca, an advisory group wrote a final report which was submitted to the environment ministers of all three NAFTA signatory countries. The report argues that transgenic corn is unacceptable in Mexico largely because of social and cultural reasons rather than known risks connected to gene flow and human health (Commission for Environmental Cooperation 2004).

The report's recommendations include reducing corn imports, labeling transgenic corn, maintaining the moratorium on commercial cultivation, mobilizing public support for in situ conservation, establishing a system to monitor the transgenic contamination, and milling GM corn as soon as it enters the country (Commission for Environmental Cooperation 2004, 31). This last recommendation is unfeasible, because of cost and because milling renders grain more susceptible to certain insects. The United States was explicit in its criticism of the report, which it called "fundamentally flawed" because its recommendations did not correspond with the CEC's scientific findings that "gene flow in and of itself does not pose risks to biodiversity and that transgenic maize varieties are no more likely to affect the genetic diversity of landraces" than other modern varieties (U.S. government comments, Commission for Environmental Cooperation 2004, Appendix, 23 July 2004).

Although the report carries no legal force and was criticized in different ways by the governments of Canada, Mexico, and the United States for basing some of its recommendations on social and cultural factors rather than exclusively on the scientific evaluation of harm to biological diversity and human health, many of the scientists and activists I interviewed felt that the process of producing the report had been important. It provided the first public forum in Oaxaca to hear the concerns of activists, NGOs, and peasants, it brought media attention to the issue, and it produced a document which recognizes that maize has cultural importance in Mexico in a way that it does not in Canada or the United States. María Colín from Greenpeace Mexico referred to the CEC as having a moral legitimacy: "The CEC study has moral weight and the report will always be used as a reference here in Mexico. I think many things have been influenced by the experience and the fact that the government hasn't taken the report seriously. For instance, scientists and academics have worked more closely with NGOs [on GM corn]. Together, they've just produced a Manifesto for the Protection of Mexican Maize" (25 July 2006). Beyond the CEC report, interviewees also felt that In Defense of Maize had brought about some important changes in the regulation of GM corn and in Mexican society. Activists and scientists cited the successes of the In Defense of Maize campaign: raising awareness about the issue of GM corn and the problems facing in situ diversity; including participants from various sectors of society, such as indigenous and peasant representatives and communities; and inducing several communities and states to seek legal recognition as GMO-free zones. In early 2009 several NGOs filed a second complaint to the CEC requesting the commission to investigate 180 acres of illegally planted transgenic maize found in the state of Chihuahua.[28]

RESPONSE AND REGULATION

After much debate and controversy Mexico finally passed a law on biosafety and genetically modified organisms (LBOGM) in 2005. The law establishes that a special regimen is needed to protect maize and other plants for which Mexico is the center of origin. There is nothing in the law itself about this regimen. Rather, the law requires that the ministries of agriculture and the environment establish the necessary measures to protect the species with the help of other state agencies. Doing so requires determining which regions of Mexico are centers of origin and centers of diversity of maize, so that GM field trials and the commercial release of the species in question can be prohibited or approved case by case. Among the criticisms of the law is that the boundaries of centers of origin and diversity are imprecise and difficult to define, but even more importantly, that they are permeable. As the economist and public intellectual Alejandro Nadal noted at a forum at Casa Lamm in Mexico City (2 May 2005), if the commercial release of a GM crop is approved, how would it be contained so as not to enter a center of origin or a GMO-free zone? Among other concerns articulated by critics was that the law does not adequately incorporate the precautionary principle or emphasize biosafety but rather promotes biotechnology; hence the nickname "Monsanto's Law" (Grupo de Estudios Ambientales et al. 2006). In terms of labeling GMOs, the law combines the precautionary principle favored in Europe with the principle of substantial equivalence used in the United States. However, the United States approach prevails in that labeling is only required when GM foods are considered "significantly different" from conventional foods (Pechlaner and Otero 2008, 364). In terms of GM imports, the law requires that the imports be labeled as intended for either human or animal consumption, and allocates further decisions about labeling to the various ministries involved (ibid.).

Since the law took effect the Ministry of Agriculture has approved field trials of transgenic maize and then reversed itself. Requests for trials were submitted to the government as part of the Proyecto

Maestro de Maíz, a partnership between public research institutes in Mexico and several large agro-corporations. The project was to involve experiments of GM corn to evaluate the possible effects on the environment and the potential benefits of GM varieties in Mexico. The ministry approved tests in the northern states of Sinaloa, Sonora, and Tamaulipas by Hybrid Pioneer, Monsanto, and Dow Agrosciences but quickly rescinded its decision.[29] Environmentalist groups argued that the permits could not be approved when neither the Biosafety Law's Special Regimen for Maize nor the country's centers of maize biological diversity had been established.[30] A group of northern producers' associations published an open letter to the president of Mexico and to the Ministry of Agriculture arguing that the Maestro project's field trials on GM maize should be approved quickly. The letter focuses on the issue of yields and ends with the phrase "More corn, more country."

In response to problems with the Biosafety Law and inaction on some of the CEC report's recommendations, a group of concerned scientists and NGOs (notably GEA) held a press conference to announce their Manifesto for the Protection of Mexican Maize (25 July 2006). The manifesto, developed during a workshop on the Biosafety Law in the Chamber of Deputies, includes seven steps to protect maize and to comply with the Special Regimen for the Protection of Maize set out in the law:[31]

Manifesto for the Protection of Mexican Maize
July 25 2006
—We declare that the Regimen for the Special Protection of Maize be considered an issue of national security which follows the precautionary principle and protects the in situ and ex situ conservation of germplasm
—that the Regimen consider the perspectives and opinions of rural and indigenous communities, academics, scientists, producers and the different levels of government
—to adopt the recommendations of biosafety studies, including that of the CEC's "Maize and Biodiversity: The Effects of Transgenic Corn in Mexico"
—to assure and designate funds to undertake a full and independent monitoring of the state of contamination among maize varieties and to remedy any contamination
—to permanently prohibit the development of non-edible industrial

products, vaccines, and experimental proteins for therapeutic use which use corn as a system of expression, and to ensure that viable materials along these lines are not allowed to enter Mexico

—to ensure that there are processes and mechanisms to bring relevant information about GM corn to all sectors of Mexican society

—to reinstitute the moratorium on the cultivation of GM corn in the open countryside for the time being

For members of In Defense of Maize the manifesto represented an important strengthening of the alliance between concerned scientists and NGOs, even if the achievement turns out to be temporary.

The ministries of agriculture and the environment determined that the northern states of Sinoloa, Sonora, and Tamaulipas were not centers of origin for maize or its wild relatives in late 2006, despite arguments that all of Mexico is in effect a center of origin and diversity for maize.[32] This controversial conclusion did not lead to field trials of transgenic maize in the north, and the Proyecto Maestro was canceled, in part because neither the regulations for authorizing GMOs nor the Special Regimen for Maize had yet been established.[33]

In 2009 the first two permits were granted for the experimental cultivation of transgenic maize in the north of the country, after a presidential decree had modified the Biosafety Law earlier in the year, making such cultivation possible for the first time since the moratorium of 1998. According to critics, this action opened the door for the deregulation of transgenic maize cultivation.[34]

On the one hand, problems remain with the regulation of transgenic maize, along with the underlying policy orientation of the neoliberal corn regime. Mexico continues to import millions of tons of maize annually (close to eleven million in 2008). On the other hand, the benefits of transgenic maize in the Mexican context are contested by the anti-GM corn campaign and its network of concerned scientists, academics, government officials, NGOs, and activists. This coalition has played the role of GM policy watchdog, influencing regulatory policy and raising awareness of the controversy in urban and rural Mexico, as well as through regional and transnational networks.

In response to the request from NGOs and peasant and indigenous groups in Oaxaca to investigate, the CEC assessed the risks and benefits of transgenic maize in Mexico, held public forums enabling activists, smallholder farmers, and other concerned parties to voice their opinions, and published a report with recommendations. The Biosafety Law recognizes maize as a special crop in Mexico, its center of origin and diversity, and provides that decisions about GM corn should take this biodiversity into account, but it also contains flaws and has been difficult to apply.

The debate and political maneuvering over how best to regulate transgenic corn imports, field trials, and commercial release continue. And there remains the wider problem of how to preserve in situ maize biodiversity when small producers face increasing impoverishment and outmigration.

The acrimonious debate over transgenic maize has involved efforts to recast expertise about maize. The Ministry of Agriculture and the biotech industry define expertise as knowledge derived from the scientific evaluation of the risks associated with gene flow. This definition both devalues non-gene-flow-related concerns and portrays gene flow between transgenes and criollos as natural, inevitable, and beneficial to the biodiversity of maize. When the scientists concerned and vocal about transgenic maize were portrayed as ideologically committed activists, and the methods that they used to identify transgenes in maize samples as sloppy, In Defense of Maize and its allies came to their defense in national newspapers like *La Jornada* and in online sites and list-serves.

The members of In Defense of Maize point to a lack of sufficient scientific studies on gene flow in the Mexican context. They also use the discourse of risk to challenge how industry and the government emphasize gene flow over other concerns, and privilege scientific expertise in assessing risk. Enrolling a broad spectrum of supporters in its efforts both at home and abroad, In Defense of Maize recasts the question of GM corn imports and regulation in terms of cultural values, peasant expertise, and corporate-led globalization. The coalition draws our attention to how small-scale farmers should be recognized as maize experts, and to the plight of these farmers under trade liberalization and the increased corporate control of agriculture. The coalition has enlisted corn as a unifying symbol for issues and concerns beyond the question of GM regulation precisely because it is such a powerful representation of Mexican culture. We now turn to rural policy and official ideas about maize production, and to how the coalition gives priority to food sovereignty and quality over market efficiency.

Chapter 2

CORN AND
THE HYBRID NATION

Without corn, there is no country (Sin maíz, no hay país)
—Slogan of the anti-GM corn, pro-maize movement

Although the cultivation of maize represents half of the area used
for agriculture in our country, we have one of the lowest average
yields in the world. To deny us possible access to new alternatives
in production, like genetically improved maize, is to condemn our
fields to backwardness and marginalization. We, the agricultural
producers of Mexico, deserve better conditions to raise our quality
of life and the well being of our families. More corn, more country.
—Letter from northern producers and producers' organizations
requesting GM corn field trials, published in national newspapers
8 December 2005

The neoliberal corn regime promotes conventional agriculture.
More is better, no matter how the food is produced, as suggested by
the letter quoted above and the slogan "more corn, more country."
This letter was sent to Mexican newspapers by northern producers
requesting that the government allow transgenic corn field trials. It
is a direct response to the anti-GM corn campaign slogan "Without
corn, there is no country," which connects smallholder maize agriculture to biodiversity and Mexican culture, rather than to "backwardness and marginalization."

The neoliberal corn regime promotes food security through imports. Under this policy regime food security is defined in relation to

the quantity of food available on the market rather than in relation to self-sufficiency at the community, regional, or national level: it does not matter where the food is produced. In recent years the tortilla and corn flour industries have turned to yellow maize imported from the United States or purchased from northern producers to make their products, in the process undercutting small-scale Mexican cultivators of white maize.

Drawing on the Mexican and transnational food and peasant rights movements, the anti-GM coalition In Defense of Maize importantly challenges this perspective and highlights the goals of food sovereignty and quality. Members of the coalition argue that Mexico should promote self-sufficiency in maize production rather than rely on imports, and that both the quality of corn available to consumers and the livelihood of maize producers matter. It does make a difference where and how corn is produced. While some coalition members oppose GM maize in Mexico but not the use of genetic engineering for agriculture more generally, others view this technology as part of the growing trend toward corporate agriculture.

In the debate over transgenic corn, competing perspectives use and reformulate earlier symbolic associations between maize and Mexico while also drawing on (and contributing to) discourses employed by transnational networks and movements. Although In Defense of Maize promotes a multidimensional analysis of GM corn imports and cultivation, some voices reproduce peasant essentialism, something that they share with the official position. The purpose of this chapter is to examine the current narratives about corn and culture in relation to the history of rural policy and debate. To understand what is at stake in the GM corn debates and how the various participants frame the issues, we need to consider the recent history of rural policy.

The twentieth century saw several important shifts in state rural policy and in the underlying assumptions about maize as both food and crop: the emergence of indigenist policy during the postrevolutionary period (1920–40), the height of agrarian reform and the consolidation of the campesino as a political identity during the presidency of Lázaro Cárdenas (1934–40), the economic "miracle" and the Green Revolution (1940s–1960s), the expansion of state-owned food and agriculture enterprises, or "the parastatal sector" (1970s), and neoliberal reform (1982–). For much of the century there was a move toward emphasizing the needs of conventional agriculture, reflecting what James Scott has called the "high modernist" model. And while the close of the century saw a neoliberal reorientation of rural and agricultural policies, key elements of the high modernist agricultural model remain, such as a focus on productivity through conventional agriculture regardless of the environmental or social impacts, and the involvement of agribusiness in various stages of production, including the provision of agricultural inputs like seed.

In official circles maize-based foods went from having negative connotations, because of their early associations with Indians and subsistence production, to becoming a complicated symbol of the hybrid nation. Maize agriculture, on the other hand, largely retained its associations with Indians and subsistence production. In the current maize debates corn is a multivalent symbol. In one register maize agriculture remains a symbol of inefficiency and a barrier to rural development. In another it represents the hybrid nation, but now as a failed or contested project of homogenization. Maize stands for the cultural alterity of the countryside and Mexico is seen as a hybrid—as in multicultural—nation. Maize also represents the key to food sovereignty in Mexico, and along with it a critique of neoliberal globalization.

CORN AND THE HYBRID NATION

Maize has long been a marker of race, class, and gender in contrast to its colonial competitor, wheat. Although foreign and domestic representations of Mexico during the colonial period included images of corn and corn-based foods (Solís 1998), wheat and wheat flour, originally introduced by Spaniards, were associated with Europe, whiteness, and civilization. In the Spanish Americas, Spaniards, or *peninsulares*, were a dominant caste with religious, civil, and military authority over other racial and legal categories of people such as blacks, mestizos (the descendants of European and Indian parentage), mulattos (the descendants of blacks and Indians), and "Indians," the original inhabitants of what Columbus believed to be the Indies. The caste system was an institutionalized expression of social stratification based on descent or perceived racial heritage. Spaniards born in the Americas, "criollos," belonged to the dominant caste and were known along with the peninsulares as "people of reason" (*gente de razón*), but their birth in the colonies was seen as negatively affecting or degenerating their character and physiognomy (Lomnitz 2001).

Under the colonial system the consumption of maize or wheat was an index of one's social status (Alonso 1995, 67). In the Tehuacán Valley wheat was cultivated by indigenous peasants and sharecroppers, while gente de razón were not expected to work with their hands. In the twentieth century agricultural labor was associated with being an Indian in the region (Henao 1980, 87). Even today younger valley residents negatively associate the agricultural labor of their parents and grandparents with Indianness (see chapter 5).

In the colonial world and newly independent Latin America, the concept of *mestizaje* referred to the problem of miscegenation and related to a shift from monogenist to polygenist theory. Polygenists believed that races were separate species, and therefore that mating across the races would either produce no offspring or racial "hybrids" who were infertile. So when mixed-raced peoples increased in number, their existence "offered an obvious challenge to polygenic presumption" (Goldberg 2002, 25).

As the mestizo population grew and was used as evidence to contradict polygenist theory, the meaning of hybridity itself underwent a shift. Hybridity came to refer to the mixing of cultural differences while simultaneously continuing to connote originally separate and unrelated peoples. "Along with this shift away from physicalist-based notions, the concept of hybridity began in turn increasingly to assume reified culturalist expression" (Goldberg 2002, 26). Even in discussions of Indians and mestizos today we can find evidence of these conceptual slippages between biology and culture, race and ethnicity.

With independence and liberal reforms the legal mechanisms of the caste system were overturned in Mexico, enabling marginalized groups to make claims as citizens rather than as members of a caste with differential access to property and power. But this process entailed the loss of special rights for Indians, and some colonial social hierarchies were maintained de facto (Knight 1990, 72; Lomnitz 2001, 48).

Mexico is often portrayed as a hybrid nation, a mix of Spanish and indigenous cultures and somatic features. The human hybrid, the mestizo, is a concept steeped in the history of the colonial project to maintain difference and distance between castes, and later the modern, post-revolutionary project of progress and development through acculturation. Maize is a symbol of Mexico and this complex history.

Maize is an apt symbol for the mestizo or hybrid nation not only because of the crop's cultural, economic, and environmental importance in Mexico but because it is a cross-pollinating plant. Pollen from one corn stalk does not usually fertilize the same plant but is carried by movement or a breeze to the silk of a different corn plant. When two races of corn are grown in proximity to each other, there is a degree of interracial hybridization (Mangelsdorf 1974). A cross-breed or hybrid plant is a "plant resulting from a cross between parents that are genetically unlike; more commonly, in descriptive taxonomy, the offspring of two different species" (Herren and Donahue 1991, 234).

From Liberalism to Revolution

In the case of Mexico . . . nationalism was built not on the
culture of bourgeoisie or of the urban proletariat, but rather
around the romanticized figure of the Indian and the peasant.
—Claudio Lomnitz 2001, 191

In liberal Latin America of the late nineteenth century mestizaje generated considerable discussion and debate. It became part of the liberal project of progress. During La Reforma (1854–76) in Mexico, the liberal government implemented legal reforms that abolished the privileges of the church and the military, allowed for the privatization of communal and corporate property, and removed legal distinctions between Indians and other Mexican citizens. Liberals viewed free trade as the proper path toward economic development and opposed the power of the church. In the Tehuacán Valley these reforms enabled haciendas to buy up the lands and waters of Indian villages, as discussed in chapter 3.

At a national level these reforms generated a conservative backlash. The conservatives supported the church and economic protectionism in order to encourage industrialization. A civil war broke out and was won by the liberals under the leadership of Benito Juárez. The liberal leader Porfirio Díaz became president of Mexico in 1876 and remained in power for thirty-five years (with the exception of the years 1880–84). Under his rule, known as the "Porfiriato," he undertook an ambitious liberal modernization program to construct a centralized political system and economic infrastructure. The government encouraged foreign investment and the building of railroads, developed both a national market and a banking system, and expanded mining centers, haciendas, and commercial agriculture. State functions were centralized, and a small cadre of officials (*científicos*) who promoted modernization policies rose to political dominance. Although the Porfiriato was a centralized and authoritarian model of government, it was mediated by regional and local political configurations. The state of Puebla was a particularly Catholic and conservative area, owing in part to the role of haciendas in its center and south (including the Tehuacán Valley) and in part to the interpenetration of civil and religious authorities (Vaughan 1997, 49). As a result, even though there was a supporter of Díaz in the office of governor (Mucio P. Martínez, 1892–1911), the official liberal culture of secularism was not very effective. Porfirian modernization was effective, however, in reviving industry in Puebla with the establishment of new textile factories (1892–1902).

Food had been linked to racial hierarchies in Mexico during the colonial period, an association that continued during the Porfiriato. Senator Francisco Bulnes argued that of the three "races" of man—the people of wheat, rice, and corn—it was the people of wheat who were truly progressive (Pilcher 1998, 77). As the food historian Jeffrey Pilcher (1998, 78) has explained, "The language of proteins and carbohydrates offered a scientific explanation of Indian underdevelopment that did not resort to the racist, deterministic doctrines of Social Darwinism. Mestizo *científicos* rejected the idea that their Indian ancestors were inherently unfit and embraced instead the notion that maize had oppressed pre-Columbian peoples. Salvation therefore lay in the adoption of European culture, especially the consumption of wheat bread."

Corn lay at the heart of rural and indigenous backwardness, and wheat was seen as a key to salvation. This popular theory offered the possibility of development and progress for the Indian masses and the Mexican nation as a whole, since it was the production and consumption of maize that held indigenous peoples back, not an innate biological inferiority. Yet at the same time, the state-building project of Porfirian liberalism was influenced by a belief in European superiority and social Darwinism. It sought the forced assimilation of Indians into mainstream society and waged large-scale wars against the Maya in the Yucatán and the Yaqui in Sonora (Knight 1990).

Indigenismo

After the revolution (1911–20) indigenist state policy, or indigenismo, sought to foment a shared national culture through the education and cultural assimilation of indigenous peoples, who were seen as culturally distinct. The founders of indigenismo in Mexico, such as the anthropologist Manuel Gamio, stressed the validity of all cultures and the equality of all races (Lomnitz 2001, 53). This was quite a departure from Porfirian rule. But non-Indian intellectuals also viewed the ethnic heterogeneity of Mexico as a barrier to progress. Anthropological and archeological studies gathered information about indigenous customs and practices in part to serve the cause of Indian assimilation.

State officials and anthropologists identified Indians primarily through social and cultural practices such as food, dress, language, social organization, hygiene, and agricultural technology, rather than through somatic features. Smallholder corn production and consumption were defining features of Indian ethnicity, and beginning in 1940 the official census included tortilla consumption as a sign of poverty (Pilcher 1998, 91). Yet although Indianness was officially defined through social and cultural characteristics (including signs of socioeconomic status like tortilla consumption), biological heredity was often reinscribed on social and cultural difference. Race was used "as a common but genetically unsound shorthand for ethnicity" (Knight 1990, 75).

The indigenismo of revolutionary leaders and the early post-revolutionary state was underpinned by contradictions and tensions. It promoted the assimilation of Indians, but it simultaneously worked to reinforce ethnic differences, marking Indians as culturally distinct and subordinate (Friedlander 1975). In addition, Indians were seen as capable of individual and collective change, but if they changed (say, by no longer speaking an indigenous language, migrating to an urban area, or gaining a formal education) they were no longer seen as Indians but rather as mestizos (Martínez Novo 2006, 7). Culture and ethnicity, although assumed to be changeable and capable of being learned, often fell back on a notion of biological race and were essentialized—treated as innate and unchanging. This meant that although it was possible for individuals or groups to change their ethnicity (to become mestizo), the cultural traits of an entire ethnic group (Indians) were relatively static. Similar opinions about Indianness are found, as well as reformulated, in the Tehuacán Valley today.

Post-revolutionary indigenismo was distinct from the policies of the Porfiriato in several ways: for one, it viewed the state as the protector of the disenfranchised Indian (Knight 1990, 78). It also strove to improve indigenous people's standard of living through public education, the creation of an institutional infrastructure, and agricultural programs (Gutiérrez 1999, 91). But at the same time the two ideological positions had some similarities. In the early post-revolutionary period the nutritional theory of Indian backwardness remained popular (Pilcher 1998). For instance, José Vasconcelos, author of the well-known post-revolutionary nationalist tract *The Cosmic Race*, which represented the mestizo as an advanced race, argued that Mexico would not develop as a nation until wheat replaced corn (Pilcher 1998, 91). Vasconcelos proposed the establishment of the House of the People (Casa del Pueblo, 1923) and the Cultural Missions (Misiones Culturales, 1925) to help educate Indians in more modern agricultural methods and the ways of western civilization. It is notable that although indigenist intellectuals and officials challenged racism and highlighted Indianness as a set of cultural practices that are mutable, learnable, and equally valid, a framework of biological race and hierarchy informed their perspective (Knight 1990, Martínez Novo 2006). This indigenismo reclaimed indigenous folklore and customs as part of its nation-building project, but it also promoted acculturation through racial and cultural mixing, or mestizaje. Mestizaje was seen as a means of development, with the mestizo as the modern citizen. As the historian Alan Knight explains, "according to the emerging orthodoxy of the Revolution, the old Indian/European thesis/antithesis had now given rise to a higher synthesis, the mestizo, who was neither Indian nor European, but quintessentially Mexican" (1990, 84). Hybridization, or mestizaje, was progress: the Indian and the European intermixed to become a new, better, and internally homogeneous category, the mestizo.

From this period until 1968 much of Mexican anthropology was linked to indigenismo. Anthropology investigated the ways rural, indigenous Mexico was excluded from national development, often overlooking rural mestizo populations or the impact of state-directed reforms (Hewitt de Alcántara 1984, 121). The investigations focused on how rural Mexico was a separate world of tradition, connected to modern mestizo society through the (often exploitative) role of the cultural broker, market intermediary, or political boss (cacique). Tradition and modernity had racial connotations in Mexico, as well as spatial ones, the rural and urban.

Bastards of Colonialism

The anthropologist Arturo Warman has argued that maize is a metaphor for the Mexican nation as a bastard child of colonialism (2003 [1988]). As indicated by the popular Mexican saying "Somos hijos de la chingada"—which can be interpreted to mean "We are the children of La Malinche," "We are bastards," or "We are screwed"— the mestizo and indeed the nation are viewed as the product of Spanish colonization, the progeny of betrayal and violence. La Chingada refers to La Malinche, the Indian interpreter and mistress of the Spanish conquistador Hernán Cortés. Together they had a son, Don Martín Cortés. La Malinche is thus considered a symbol of betrayal and domination, and simultaneously as the mother of Mexican culture (Paz 1961).

In this way the post-revolutionary vision of the mestizo or hybrid nation also encapsulates the sexual violence and social transgression of its inception. Spanish conquest was the origin of the Mexican mestizo nation, and indigenous peoples were seen as the violated and dominated part of the equation; they were, in other words, feminized. In Mexico relations of exploitation and submission are often charged with gendered and sexualized meaning. Gender is constitutive of social relations, but it is also a way of representing relations of power. To be considered an Indian, part of an exploited or dominated social identity in Mexico, often implies feminization and dehumanization (Alonso 1995; Lomnitz 2001; Paz 1961). The nationalist discourse of hybridity was based on structural violence in other ways as well: the social and economic domination of indigenous peoples continued in many regions of post-revolutionary Mexico, and progress and development were promoted, at least initially, through the acculturation of Indians.

Thus in another register corn represents the possibility of progress through violence, the dialectic of a gendered and racialized modernity. Maize is associated with femininity through its designation as a crop of indigenous origin—the violated part of the hybrid equation—and it is also a symbol of fertility and life in Mayan and Nahua myths, and a food associated with women's work in the domestic sphere. Corn-based foods are associated with femininity because they occupy many hours of women's domestic work in rural areas like the Tehuacán Valley, where corn tortillas are still made by hand. Maize-based foods and agriculture, like women's reproductive work generally, are a marker of the "traditional" and the feminine in that they constitute part of the domestic. In the southern Tehuacán Valley, as in many other places, the daily practice of corn production, processing, cooking, and consumption delineates gender roles and social spaces. In the valley women are responsible for the crop's preparation as food and the cornfield is considered a male domain, but in other rural areas women have become more directly engaged in milpa production because of male labor migration (Rimarachín, Zapata Martelo, and Vázquez García 2001).

Twentieth-century improvements to the *nixtamal* mill for tortilla dough (*masa*) and the mechanization of tortilla production, first in urban areas and eventually in many rural areas, gradually loosened the association of tortillas with femininity, although never completely (Gutmann 1996; Pilcher 1998). In San José making tortillas remains women's work, although the first corn mill arrived in town in 1953, when according to one interviewed resident, "the whole town would line up to have their corn ground" (interview, 2 July 2002).

Agrarian Reform and the Campesino Identity

Agrarian activists of the revolution and post-revolutionary officials portrayed the campesino as the legitimate heir to the land usurped by haciendas and the heart of Mexican cultural heritage (Boyer 2003). As a broad category inclusive of rural people from different circumstances and backgrounds—indigenous people, mestizos, villagers, subsistence farmers, landless day laborers—the campesinos served to legitimize the post-revolutionary state in its claim to represent "the masses." As Boyer (2003, 23) explains, "campesino identity, as postrevolutionary leaders and agrarian activists originally conceived of it, was not really meant as an accurate description of rural people's exact relationship to the means of production or their ethnicity, much less as a reflection of the way that rural people described their own place in society. Instead, campesino identity originated as a political category. Postrevolutionary state makers elaborated a form of revolutionary populism that defined Mexican society in broad, essentially political terms. By the time Cárdenas became president in 1934, the postrevolutionary state was claiming to act 'in the name of the masses' (that is, on behalf of workers and campesinos), as historian Arnaldo Córdova has shown." For many rural peoples this identity resonated with their lived experience or collective memory of disenfranchisement and exploitation. There were of course rural peoples and communities who did not identify with the campesino, as suggested by Nugent's and Alonso's research (mentioned below).

Under the revolutionary constitution (1917) campesinos were granted economic and political rights, primarily through the *ejido* land grant, which combined communal title vested in the state with either communal or individual usufruct rights to work the land.[1] Article 27 of the constitution enabled the state to dispense property and land in the "public interest"; it gave towns and villages the right to petition for land; it deemed returnable the land that had been expropriated under the Porfiriato; and it allowed only Mexican-born citizens to own land. Under the presidency of Lázaro Cárdenas (1934–40) more land was distributed to peasants than in years previous or subsequent, and the ejido became a permanent form of land tenure. The ejido often generated interfamily tensions, since it was the head of the household who was granted land rights (de Janvry, Gordillo de Anda, and Sadoulet 1997, 9). Additionally, this was a gendered process, as access to ejidos and communal lands was largely granted to male heads of households.

Post-revolutionary agrarian reforms helped institutionalize the campesino as a state-recognized political identity. These reforms recognized community claims to land and water (and in some cases the formation of communities) based on participation in the revolution and the need to create and fortify political loyalties. "The Mexican *ejido* was conceived as a compromise to serve simultaneously as an instrument of political control, a means for the organization of production, and a body of peasant representation" (de Janvry, Gordillo de Anda, and Sadoulet 1997, 1). But agrarian reform also varied enormously by region. As Nugent and Alonso (1994) have shown, some communities petitioned the post-revolutionary state to have their expropriated communal lands restored or recognized as ancestral or indigenous rather than accept the government's "gifts" of land grants and the associated interventions into rural communities by the agrarian bureaucracy. In some areas water was as important a resource of agrarian reform as land (Aboites Aguilar 1998). This was so in the semi-arid Tehuacán Valley, where Porfirian-era haciendas had usurped access to spring water. Today, despite recent privatization efforts in the valley, the ejido remains a form of land tenure, as it influences access to state resources and informs local campesino identities.

The post-revolutionary state also generated and reproduced ethnic differences and identities despite the early indigenist goal of assimilating Indians. Cárdenas implemented indigenist programs through socialist education in the countryside and established the Councils of Indigenous Peoples, which were affiliated with the ruling party (PRI) and served as another avenue for state corporatism (Martínez Novo 2006, 65). Indigenous communities and individuals took up and employed official discourses about ethnic identity in selective and sometimes conflicting ways.

Although Cárdenas championed social redistribution policies, he had to appease conservative state governors and their regional allies, who were more interested in securing their own power than in land redistribution. The president tried to do so by reforming the national party in a way that allowed for the national representation and cooptation of workers and peasants. The executive committees of the ejidos were linked to the state through several institutions, including the regional peasant committees, the state league of agrarian committees, and the National Executive Committee of the National Confederation of Peasants (or CNC, which represents the peasant sector in the PRI). The state oversaw the channeling of public funds to the ejidos through an agricultural lending agency, which became an avenue for political corruption and the manipulation of rural producers.

Cárdenas's administration also set up a system to purchase maize from rural producers at a support price and resell the grain in other areas. This was the beginning of what would later emerge as one of the largest state enterprises in Mexico, the basic staples agency CONASUPO. At the same time the administration's agricultural policies sometimes worked at cross-purposes. One long-running agricultural education project that had started in the 1920s and continued in the 1930s was the Pro-Maize Campaign (Campaña en pro del maíz), which was created to persuade peasants through popular media, rural lectures, and demonstrations to select and test seed according to certain morphological characteristics and to properly store and disinfect seed before sowing (Cotter 1994a, 78–80). Cárdenas also developed projects to alter Indian diets. Some of these projects encouraged reducing the production and consumption of corn in favor of what were thought to be more nutritional crops, like wheat and soybeans (Pilcher 1998, 92, 110). Despite the administration's pro-indigenous discourse, the nutritional theory of Indian backwardness continued to influence public health policy and rural development projects. As part of Cárdenas's socialist education program, rural teachers taught modern hygiene and agriculture, and promoted domesticity, reduced alcohol consumption, and secularism. Rural producers were encouraged to take up "modern" agriculture, to diversify crops, to sell their crops on the market, and to adopt new methods of fertilization and proper seed selection (Vaughan 1997, 85).

Changes in State Policy

The goals and consequences of indigenismo were at times disputed and contradictory. An indigenist policy that promoted indigenous acculturation or incorporation into the dominant mestizo culture was not the only school of thought. Another current of indigenismo emphasized the unique cultural contribution of indigenous peoples, arguing that some features of their culture should be maintained. Moisés Sáenz, for instance, promoted teachers as cultural missionaries who could help to incorporate indigenous communities into the wider national culture; but later, as vice-president of education, he began to question elements of this incorporationism (Hewitt de Alcántara 1984, 15). Although some state officials increasingly put an emphasis on retaining what they saw as the positive aspects of indigenous culture, assimilation and mestizaje remained an official goal (Gutiérrez 1999, 97–8).

The National Indigenist Institute (Instituto nacional indigenista, or INI), established in 1948, was informed by these different and at times contradictory elements and ideas of indigenismo. The establishment of INI's rural centers (Centros coordinadores) has been described as both a way of bringing state institutions to isolated indigenous communities and a way of further isolating the communities by confining indigenous issues to the institute, away from the central state apparatus (Hindley 1996, 227). INI's projects met with varying degrees of success. Judith Friedlander (1975), who studied the effects of the cultural missions in a Nahua community of Hueyapan, Morelos, found that indigenist policy had successfully discouraged certain indigenous customs, and that when a person acquired the material symbols of well-off urban mestizos, Hueya-peños considered that person less Indian (1975, 131). In other regions indigenist policy was unsuccessful or faced outright resistance (Gutiérrez 1999; Vaughan 1997).

Other state policies also had uneven and regionally varied consequences. In Puebla the socialist education program forwarded by Cárdenas was never fully supported or promoted by the state governor. The state of Puebla had been rife with political conflict since the revolution (Pansters 1990, 49). Nineteen governors ruled the state within a ten-year period. From this power vacuum General Maximino Ávila Camacho emerged as a political leader, becoming governor in 1937. His influence lasted thirty-five years. His supporters, the *avilacamachistas*, had close relations with the conservative bourgeoisie and the church. As governor he was responsible for reorganizing the military and founding the paramilitary group *guardias blancas* in an attempt to calm the political turmoil and conflict between peasants, political bosses, landowners, and roaming bandits (Pansters 1990, 59). Cárdenas supported the authoritarian and conservative avilacamachistas.

During and after the revolution the residents of the Tehuacán Valley were often the victims of bandits and rival political factions. Competing claims to political power and the state distribution of spring water helped to redraw community alliances and shaped the regional reputation of San José as a violent Indian town. As residents fought over water claims, they negotiated and redeployed stereotypes circulating in the region about the violent nature of Indians. The mestizo teachers who arrived in San José Miahuatlán for the first time during the period of socialist education in the 1930s may also have been a source of discourses about Indian ignorance and violence.

To summarize, both the liberalism of the Porfirian dictatorship and post-revolutionary indigenismo heralded mestizaje or hybridity as a model of the nation, albeit in different ways. In both periods corn was central to the state's racialized idiom of Indianness, and the crop was the target of various efforts to displace or modernize indigenous ways of life. Although post-revolutionary indigenist policy hoped to improve the lives of indigenous Mexicans, in some cases it also helped to reproduce prejudices about Indian backwardness.

The emergence of indigenous movements in the 1960s helped shift the orientation of indigenismo (Gutiérrez 1999, 90). And after the political crisis of 1968, which followed the government massacre of protesting students in Mexico City (on 2 October 1968), indigenous communities increasingly criticized indigenist policy and the Mexican state for exploiting rural peoples. The INI turned away from its goal of assimilation and promoted indigenous "participation." At the same time, the state sought to repress or co-opt autonomous local and regional indigenous mobilizations (Hindley 1996, 228).

By the 1980s the plight of indigenous peoples was such that the call for collective rights was growing louder in Mexico and was debated in international forums such as the United Nations. Mexican anthropologists like Guillermo Bonfil Batalla (1996 [1987]) argued that mestizaje was an unsuccessful ideology. For Bonfil authentic Mexico had its roots in Mesoamerican Indian civilization, and he believed that the state should recognize this by adopting policies rooted in the principle of ethnic pluralism.

MODERNIZING AGRICULTURE

While mestizaje was gaining ground as the nationalist project of cultural and racial homogenization through mixing, the post-revolutionary state attempted to modernize agriculture. In the 1940s the Rockefeller Foundation worked in collaboration with the Mexican government to improve wheat and corn varieties, among other crops. It was the program's wheat varieties that led to an increase in agricultural yields. Together with the government's construction of roads and irrigation works, as well as agricultural banks and subsidies, the new wheat varieties transformed the uncultivated areas of the north and northwest into farmland, providing food for the rapidly expanding urban sector. In contrast to wheat, the program's hybrid corn and improved, open-pollinated varieties were less successful.

During the Cárdenas years agronomists were viewed as political agents of agrarian reform and were expected to join the National Confederation of Peasants (CNC). State agronomy was "holistic" in that it considered the financial constraints faced by peasants and encouraged peasants' participation in seed improvement. Two trends developed in state agronomy after Cárdenas's administration. First, while the state had funded rural campaigns in the 1920s and 1930s to encourage seed selection among peasants in an effort to increase their corn and wheat yields, by the 1940s state programs for seed improvement privileged the expertise of agronomists in the breeding center or lab over that of peasants in the field. This reflected new sources of funding and a new, modern approach to agronomy that produced the Green Revolution.

Second was the collection and classification of different varieties of maize undertaken by the Rockefeller Foundation's Mexican Agricultural Program (MAP) in conjunction with the Autonomous University of Chapingo in the 1940s. MAP was responsible for the Green Revolution of the 1960s and provided the basis for the world's largest and most important maize seed, or germplasm, bank, now based at CIMMYT (the International Corn and Wheat Institute), INIFAP (the National Institute of Forestry, Agricultural and Livestock Research), and the Autonomous University of Chapingo. The scientific classification and naming of corn varieties drew on ethnic terms and reproduced some of the plant's ethnic and racial connotations. Eric Wellhausen, Paul Mangelsdorf, and the Mexican Efraím Hernandez Xolocotzi recognized twenty-five "more or less distinct races" of maize, which they divided into four groups: Ancient Indigenous, Pre-Columbian Exotic, Pre-historic Mestizo, and Modern Incipient. In their theory of the development of corn varieties they found evidence that the Pre-Columbian Exotic from Guatemala had mixed in Mexico with both the wild grass teosinte and ancient indigenous varieties, which were types of popcorn. Remembering their breakthrough, Mangelsdorf explained that the "almost explosive diversification" of corn races produced from this hybridization was named "Prehistoric Mestizo" because mestizo is "the Mexican word for racial hybrid" (Mangelsdorf 1974, 102). These maize scientists linked well-defined races of corn with "pure" Indians, as in this explanation by Mangelsdorf (1974, 145): "Among the pure-blooded, non-Spanish-speaking Indians of Guatemala rigid selection for a type of seed is often practiced (Anderson, 1947). This probably accounts for the fact that in Guatemala there is a high correlation between the percentage of *indígenas*, pure Indians, and the number of well-defined races of maize (Wellhausen et al., 1957)." This correlation between ethnic diversity and the biodiversity of maize is under debate in more recent studies (Perales, Benz, and Brush 2005).

Toward the end of Cárdenas's term in office in the late 1930s, shortages of basic food crops and high inflation contributed to a mounting dissatisfaction with and consideration of the role of agronomy and agrarian reform in modernizing agriculture. They culminated during a food crisis (1938–39) when the supply of cereal grains and beans became a serious problem because of transportation difficulties and crop shortfalls. As a result government expenditures on corn imports skyrocketed (Cotter 1994a, 258). While many agronomists remained committed to the goals of agrarian reform and a holistic agronomy, some began to argue that agronomic training should be separated from the state agrarian reform institutions, especially the National Confederation of Peasants (CNC), and devote more attention to experimental science (Cotter 1994a, 302–4). The food crisis helped to fuel criticism of Cárdenas's emphasis on land redistribution.

Claiming that Cárdenas's reforms were inefficient, the subsequent governments of Manuel Ávila Camacho (1940–46) and Miguel Alemán (1946–52) pursued a dual-track agricultural policy. On the one hand they continued to maintain the *ejidal* sector, which consisted largely of subsistence and smallholder agricultural production although not at the levels of Cárdenas's administration. On the other hand, Ávila Camacho and Alemán helped capital-intensive agriculture in an effort to raise productivity and export revenues, and to provide cheap food for a mushrooming urban labor force (Hewitt de Alcántara 1976). Unlike Cárdenas, who emphasized land redistribution and indigenous education, the counter-reform expanded efforts to promote the modernization and commercialization of agriculture.

High Modernist Agriculture

The shift in Mexican state policies in many ways reflects what James Scott (1998) has called "high modernist" agriculture, a one-size-fits-all model of rationalized, industrialized agriculture on a large scale. In the name of productivity and efficiency, this approach to agriculture gained popularity in the twentieth century, first in the global north and then in development programs for the global south. Modernized agriculture depends on mechanized farms, monocrops, and the use of industrially produced and capital-intensive inputs (for example, synthetic fertilizer over manure, or hybrid seed over farmer-saved seed). This type of agriculture involves agribusiness corporations in various stages of production and increases the distance that foods travel between production and consumption. Scott argues that this agricultural model is not necessarily more productive than smallholder agriculture, but that it is more efficient in terms of control (1998, 189). High modernist agriculture sets out to make rural spaces legible to the state in an effort to govern or control them more efficiently. Scott points out, though, that high modernist agricultural schemes are not always successful in meeting their goals, and that they can have unforeseen consequences.

A key characteristic of the high modernist model is that it privileges experts in the laboratory and the government office over practitioners in the field, and western scientific knowledge over "métis" or practical knowledge, which it disdains. "The logical companion to a complete faith in a quasi-industrial model of high modernist agriculture was an often explicit contempt for the practices of actual cultivators and what might be learned from them." (1998, 304). The production of knowledge for high modernist agriculture narrows the focus of the enterprise to inputs and yields, in contrast to the more holistic approach of cultivators' practical knowledge. This narrowing of vision means that outcomes other than yields (such as environmental impacts) are not considered except when they affect production. Moreover, this approach presumes that particular environmental or cultural contexts do not matter. It treats information about agriculture and crops in an abstract, technical, and context-free manner (Scott 1998).

In Mexico the shift toward catering to the needs of commercial, large-scale, and conventional agriculture reflects this high modernist model. The problems facing rural smallholders were increasingly treated as technical problems, which required the expertise of agronomists and government officials. Despite the heroic figure of the campesino in post-revolutionary state and popular representations, a strain of agricultural policy and agronomy devalued the practical knowledge of peasants. In more recent years the neoliberal corn regime has reoriented Mexican state-rural relations and agriculture; however, elements of the model of high modernist agriculture remain.

The Green Revolution

Mexico was the first developing nation to undergo an agricultural Green Revolution, which took place primarily in the country's irrigated regions of the north. The Rockefeller Foundation's research and training project, named the Mexican Agricultural Program (MAP), opened its doors in 1943 in collaboration with the Mexican Ministry of Agriculture and Animal Husbandry (now SAGARPA). MAP, along with the semiautonomous office of the ministry (OEE, or Oficina de Estudios Especiales), proposed to raise the productivity of maize and wheat, among other crops, as a means of alleviating hunger, raising the nutritional levels of the average diet, and reducing Mexican dependency on basic food imports. The Green Revolution was also designed in part to dissipate social unrest: a green revolution instead of a "red" one (Cotter 1994a; Jennings 1988).

In the 1940s the Rockefeller Foundation also funded nutritional studies in Mexico. The foundation was interested in bringing to Mexico its experience with treating pellagra (a vitamin deficiency caused by a lack of niacin) in poor communities of the southern United States, where residents ate a lot of cornmeal (Pilcher 1998, 93). This disease was rare in Mexico because of the nutritional benefits of nixtamalization, the traditional process by which tortilla dough is made, yet maize had been officially viewed as nutritionally inferior to wheat for many years. In the 1940s scientific evidence began to suggest that wheat and corn were nutritionally equivalent, and as a result public officials no longer encouraged the substitution of wheat for corn with the same fervor. Still, the consumption of wheat bread gained popularity in Mexico, and state actors continued to design projects to replace maize with ostensibly more nutritious foods like soybeans (ibid. 93, 110).

By the time the OEE ceased operations in 1961 and many of the Rockefeller scientists had moved to a new organization that would become CIMMYT, the surge in wheat production known as the Green Revolution was under way. Yet as critics of the Green Revolution have pointed out, two decades after "the initiation of the joint technical assistance program in 1943, Mexican wheat yields were the highest in Latin America, while the average yields of corn were among the lowest" (Hewitt de Alcántara 1976, 26). This project trained a generation of Mexican agronomists and professionalized the practice of agronomy in Mexico.

MAP identified hunger, malnutrition, and national food self-sufficiency as technical problems, largely separate from questions of land reform and access to resources, questions which were in effect "depoliticized" (Ferguson 1994). MAP claimed that its technology was scale-neutral, benefiting farmers large and small, but many of its attempts to combat hunger and reduce dependency on imports stressed the goals of commercial productivity (Barkin and Suárez 1983; Hewitt de Alcántara 1976). This process of depoliticizing problems or rendering them technical characterizes the high modernist model for agriculture, but it is also at work today in the GM corn debates and policies affecting the Mexican countryside.

Between 1950 and 1970 agricultural production increased more rapidly than the population, and by 1965 80 percent of all land devoted to wheat was planted with MAP varieties, while "its corn strains accounted for only a tenth of the total acreage" (Cotter 1994a, 701). The Green Revolution not only favored certain crops but benefited particular regions and farmers.[2] MAP contributed the new technologies needed in transforming the uncultivated, desert areas of the north and northwest into wheat-producing farmland. The government-financed roads and irrigation works, agricultural banks and credit, and guaranteed wheat prices were more accessible to large-scale farmers and benefited them more. Small landholdings increasingly became sources of cheap labor for expanding industries and agro-exporters—a point emphatically made during the agrarian debates of the 1970s. These policies and subsidies contributed to the postwar industrialization of Mexico, and in some areas maize was displaced for more remunerative crops. Mexico would soon have to import maize once again, as well as other important grain crops (Barkin and De Walt eds. 1990, 35).

As more profitable crops replaced maize in the north of the country and the government imported yellow corn from the United States to supply urban consumers, corn became a nostalgic symbol of mestizo middle-class nationalism. In his history of Mexican cuisine (1998), Jeffrey Pilcher demonstrates that during the 1940s, a period marked by rapid urbanization and industrialization, import substitution, and the "Mexican Miracle" of economic growth, corn-based foods that had formerly been considered poor, nutritionally deficient, and backward were appropriated by the growing middle and urban classes as "authentically" Mexican. In the same period, but increasingly from the 1960s, the diet of the poor incorporated sugar and fat, especially in the form of soft drinks and processed foods (ibid. 116). Meat consumption also rose among wealthier Mexicans, generating an increase in the production of sorghum as cattle feed, and a decline in the area devoted to maize cultivation (Barkin and De Walt eds. 1990).

The Corn-Provisioning System (CONASUPO)

At the same time that one branch of the state promoted the modernization and commercialization of agricultural production, another state agency developed and expanded a corn-provisioning system. Maize provisioning and food policy were central to state corporatism and political campaigns in a largely rural society that was rapidly urbanizing from the 1940s onward. The price and availability of corn or corn tortillas could make or break a presidential administration (Ochoa 2000).

Originally founded under Cárdenas, the state marketing agency for basic staples became the largest welfare agency of the Mexican state in the 1970s. Cárdenas's administration began to purchase grain from producers throughout the country in 1936 at a support price for resale. This agency, later called CONASUPO, constructed grain elevators in the 1940s and expanded its network of warehouses to store corn and beans in rural zones in the 1950s, becoming one of the largest state enterprises in Latin America.

In the 1970s the government more than doubled the number of parastatal agencies. CONASUPO worked to ensure that maize would be bought from small and medium-sized producers at the support price. The grain was channeled through private intermediaries and state agencies to urban consumers and the corn flour and tortilla industry. Along the way CONASUPO often stored maize in its large network of warehouses and sold to the public through its retail stores (DICONSA), managed by local cooperatives. But before changes were implemented in the 1970s most small producers could not deliver the minimum volume of a standardized quality required for drop-off at state centers. Also, retail sales were centralized in urban areas (Hewitt de Alcántara 1994, 5). This made the political and economic role of intermediaries very strong in the corn-provisioning system.

Programs were put in place to support the buying and selling of small amounts of corn and other subsidized grain. In this system of state subsidy and provisioning, consumers, subsistence producers, and commercial producers had competing interests. Oligopolistic grain markets are a widespread problem in Mexico, one that CONASUPO and DICONSA sought to ameliorate. Some researchers warned that despite the problems of inefficiency and corruption, particularly among intermediaries, the dismantling of CONASUPO and DICONSA would be ruinous for Mexican rural producers (Hewitt de Alcántara 1994). CONASUPO was nevertheless closed in the late 1990s.

In the head town of San José in the Tehuacán Valley, a CONASUPO cooperative store was opened through the efforts of a local water association and an ejido. When there was a local shortage of corn in the early 1980s, the store provided what residents considered "flavorless" yellow corn to consumers, but it never purchased corn from Sanjosepeños. When residents took up the production of elote for sale, intermediaries largely from other towns—those who owned trucks to take the product to urban markets—were key actors in selling maize on the regional market, just as they had been in the sale of valley wheat in previous years.

The Agrarian Debates

As Mexican society underwent rapid urbanization and industrialization, Mexican anthropology became part of a growing international debate about "the agrarian question." By the 1970s Mexican scholarship was immersed in an argument between the agrarian populists (campesinistas) and left-wing proletarianists (descampesinistas), a debate that can be characterized as a tension between traditionalist and modernization discourses (Kearney 1996).

Corn production was an important if not always explicitly stated feature of the agrarian question, because smallholder agriculture is dominated by maize production in Mexico. The central questions addressed in the agrarian debate were as follows: Is the "peasant economy" viable? Is smallholder, largely subsistence production disappearing or being reproduced under capitalism? And at a time when domestic maize production was falling, the populists asked, can such production supply the majority of the nation's maize requirements?[3] Although many authors adopted insights from both sides, for purposes of clarification we consider the differences between the two positions here. After more than forty years some of the same questions are being asked in the maize debates between alter-globalization populists and neoliberal technocrats.

On one side of this debate campesinistas invoked the work of A. V. Chayanov to argue that the assumptions of classical economics, particularly that of capitalist rationality, could not be applied to peasant societies, which defy the disintegrative pressures of capitalism. In the face of unfavorable conditions, peasants will intensify their unpaid labor, considered an "irrational" response under capitalism by classic economic theory. Peasants were seen as having a tendency toward "self-exploitation." They were an abundant reservoir of cheap labor, exploited through an unequal exchange with capitalists. Capitalists took advantage of peasants because rural households could fall back on "self-provisioning," which in Mexico tends to mean corn production for household consumption.

On the other side of the agrarian debates, the descampesinistas drew on the work of V. I. Lenin (1961 [1899]) and argued that transforming peasants into wage laborers was both inevitable and politically desirable. Peasants were viewed as a residual category from an earlier era on the path toward proletarianization (Bartra 1974). To a certain extent the descampesinista position shared elements of a modernization agenda (Kearney 1996). At their worst, both positions transposed to the Mexican countryside the framework developed for turn-of-the-century Russia.

The campesinista position had many variations. Adherents of one strain argued that peasants could survive indefinitely under Mexican capitalism because of the household's ability to maintain itself by relying on smallholder agricultural production and community organization. The organization of rural Mexican indigenous and peasant communities was often characterized by ritual kinship (*compadrazgo*), collective labor duties (the *faena*), and a civil and religious hierarchy of community posts (*mayordomías*). The discussion of these community mechanisms was influenced by anthropologists such as Eric Wolf and Gonzalo Aguirre Beltrán.

An early contributor to the agrarian debates, Wolf (1955) argued that the corporatism of indigenous communities was not a cultural survival but a product of and reaction to colonialism. In his oft-cited work he defined a type of "closed corporate community" as an indigenous peasantry that retains "effective control" of land (in contrast to tenants, such as those on hacienda systems); as being primarily concerned with subsistence rather than reinvestment; as practicing strong territoriality and endogamy, thus preventing the renting or usurpation of communal land by outsiders; and as belonging to communities that had mechanisms for distributing resources (1955, 458). Redistribution of resources did not, however, always work to level class differences within communities; it may in fact have strengthened inequality and the political prestige of élite members of the communities (Wolf 1986, 327).

Many of these closed corporate communities lost control over their own livelihoods when their land was forcibly removed from communal jurisdiction during the 1850s Liberalist period and the Porfiriato. This loss of control was represented by Wolf as giving rise to a second ideal type, the "open" peasant community, a community shaped through relations with the wider world market and the particular patterns of colonial settlement. Wolf's ideal types were an attempt to move away from the notion of the bounded, isolated community: to show how what seemed geographically remote and isolated from larger society was shaped through a history of interaction. Moreover, subsistence and cash crop production do not always represent different stages but may coexist within the same community. The "open" community was characterized by private and individual land ownership, the erosion of leveling devices accompanied by conspicuous shows of wealth (in *fiestas*, for example), and increased activity within the community by "outsiders" such as non-peasants and non-Indians, who sometimes acted as cultural brokers (1955, 461).

Wolf later clarified that his distinction between open and closed communities had been overly schematic, but his main point remained: the peasantry was historically formed through relationships of exploitation (1986; 1955, 454). Addressing similar questions, the anthropologist Gonzalo Aguirre Beltrán (1967) argued that modern rural exploitation had emerged through a colonial caste system that left indigenous regions on the periphery of development.

There were many problems with the campesinista perspective. Suffice it to say that the peasantry was variously seen as a "natural economy" that valued self-provisioning over profit seeking and risk taking, as embodying use-value over the value of exchange, and as isolated from and unaffected by wider, external markets and capitalist relations.

Despite the incorrect conclusion reached by some campesinistas that the peasantry would never disappear because it functioned to reproduce cheap labor for capital, their argument that peasants adapted to periods of crisis by expanding and adjusting the amount of their unpaid labor is an important insight, one that I find applicable to the Tehuacán Valley (see chapter 4). The agrarian debates rightly highlighted how peasant labor is a flexible resource (Chayanov 1966; Esteva 1983; Paré 1977; Warman 1980 [1976]). Peasant commodities are bought cheaply by local and regional distributors and companies, and national and transnational companies buy peasant crops to transport them to distant markets. In turn, peasant communities—sometimes located at great distances from urban markets—pay higher prices for goods because of transport costs. Peasants with small or medium-sized landholdings must also rely on credit from private sources like village lenders at high interest rates, or on credit provided by the state that makes them vulnerable to political coercion.

The descampesinistas, on the other hand, argued that no community mechanism was capable of slowing the process of proletarianization. Roger Bartra criticized dependency theory, which was popular at the time, for emphasizing circulation over the relations of production, and argued that capitalist transformation in Mexico had been incomplete. Because of the incompleteness of the process, he believed that further proletarianization would be beneficial to peasants and their political organizing (1974). In much descampesinista writing class was defined mechanically and political tendencies were often extrapolated from class position, as though an observer could surmise political affiliation based on observed livelihood strategies.

There were some within the debates who did not arrive at foregone conclusions, or who did not assume the inevitability of proletarianization or the function of the peasantry under capitalism, but rather questioned the concept of the "peasantry" as a category and emphasized regional and historical study (Cook and Binford 1990; Otero 1999). Additionally, research began to describe a process of semi-proletarianization, in which proletarianization and re-peasantization were taking place simultaneously in different regions, or as part of a unifying yet contradictory process. Some authors argued that peasants seasonally or generationally join the wage labor market, but that their doing so is not necessarily a sign, as the descampesinistas would have it, of the erosion of peasant production. For example, Luisa Paré (1977) suggested that many peasants were semi-proletariats salaried through agricultural credit, while Lourdes Arizpe (1978; 1981) maintained that migrant work subsidized the countryside and smallholder agriculture, rather than the other way round. Arizpe demonstrated that the life cycle of households was crucial in a pattern of "relay migration" from the Mazahua community that she studied to urban centers. Members of the younger generation migrated for work and provided a kind of capital for their agricultural parents, a process not unlike what I found in the southern Tehuacán Valley.

Arizpe was also one of the voices in the agrarian debates who challenged interpretations of Indian ethnicity as negative—as embodying an exploitative relationship, or as explainable wholly in relation to class. Some studies drew on earlier work to argue that Indian ethnicity was not just an external evaluation but rather an identity around which people could organize politically (Hewitt de Alcántara 1984, 171). These issues are still very much alive today. In some regions Indian ethnicity is indeed a positive political identity, whereas in the southern valley indigenous identity is not universally adopted or celebrated, and despite rural outmigration and changing agricultural practice, older residents cling to campesino political identities. The very historical and political baggage which makes the term "peasant" unpopular in some academic circles contributes to its continued popularity as an identity and self-label in regions of rural Mexico. As explored in chapter 5, this identity reaffirms the dignity and humanity of residents, while also harking back to a time when the state was believed to have a responsibility and commitment to small-scale rural producers.

The current GM corn debates are informed by these earlier agrarian debates, by traditionalist and modernist frameworks. Some participants in the anti-GM campaign romanticize corn as part of a millennial culture of use-values, not unlike the earlier work of some campesinistas. Some state officials, in contrast, echo an earlier modernization framework and the work of descampesinistas in arguing that the proletarianization of campesinos is desirable or inevitable. Their ideas also echo the late-nineteenth-century idea that maize production oppresses Indians. With the proliferation of rural migration in recent years, the agrarian debates have also been reinvented in debates about the role that migrant remittances play in sustaining the countryside. Discussions of transnational migration as either improving rural livelihoods or as a form of dependency on the United States economy at a high social cost rearticulate arguments of the earlier agrarian debates.

Now that we have a historical context for understanding the policy shift of the neoliberal corn regime of the 1980s, let us look more closely at the two main perspectives in the GM maize debates. The struggles over the fate of the peasantry and maize agriculture, and the narratives about them, frequently draw upon a conception of traditional Mexico and its culture of corn.

THE NEOLIBERAL CORN REGIME

From Self-sufficiency to Import Dependency

After the debt crisis of 1982 austerity measures and the restructuring of government agricultural extension services, marketing agencies, and the rural credit system exacerbated the long-standing difficulties faced by Mexican agriculture. Previous administrations had pursued national self-sufficiency in maize and made a related commitment to support small to medium-sized producers, while importing corn to meet an increase in demand (Appendini 1992; Austin and Esteva 1987; Otero 1999). Under neoliberal reforms the goal of self-sufficiency was replaced by policies that focused on providing urban consumers with access to cheap tortillas through grain imports, "which enjoyed low, subsidized international prices and could be obtained with cheap credit" (Appendini 1994, 148). Rural areas that practiced smallholder, rain-fed agriculture were classified as areas of rural poverty with "low productive potential," and needing social welfare assistance (de Teresa Ochoa 1996, 190). State agricultural programs sought to decentralize, reduce state duplication of services, and teach entrepreneurial attitudes to campesinos.

In the 1990s Article 27 of the Constitution of 1917 was amended to allow for the rental and sale of ejidos, and NAFTA was implemented as part of a strategy to restructure the agricultural sector. Under this trade agreement Mexico was to gradually open its doors to corn imports in exchange for guaranteed access to the market for horticultural products and other labor-intensive crops in Canada and the United States. The assumption was that according to neoclassical international trade theory, Mexico has a comparative advantage in producing such crops because of its surplus in labor and lower production costs. This increased reliance on corn imports from the United States was also deemed necessary to meet the demands of industry and a growing population.

When NAFTA came into effect the price of the grain fell drastically in Mexico. Between January 1994 and August 1996 it dropped 48 percent, a savings that was not passed on to consumers through lower tortilla prices (Nadal 2000b, 5). The tortilla consumer subsidy was also eliminated after more than fifty years. NAFTA established a tariff-free import quota system for corn, but since its implementation annual imports have surpassed the agreed-upon quota, and importers have not been obliged to pay the required over-quota fees. The remaining quotas and tariffs on corn imports ended in 2008 despite the efforts of several political organizations and coalitions to have them reconsidered or renegotiated.

The rhetoric around NAFTA suggested that a national market flooded with imported maize and lower prices would not adversely affect subsistence producers. Yet often subsistence producers are also petty grain sellers, who tend to sell at a disadvantage after the harvest, when there is an abundance of local and imported corn. These same producers are then obliged to purchase corn when their stored grain supply has run out and prices are higher (Nadal 2000b, 8, 24). A second argument advanced for NAFTA was that it would generate off-farm employment for noncompetitive, displaced rural producers (ibid., 24). There was dramatic growth in the export of fruits and vegetables such as tomatoes, cucumbers, avocados, and limes under NAFTA, but employment in export production did not make up for losses in other agricultural sectors. Mexico lost half a million agricultural workers from 1995 to 2005. Moreover, Mexico's import of grains increased faster than its exports (Pérez, Schlesinger, and Wise 2008). In the Tehuacán Valley, as we shall see, there was not enough sufficiently remunerative employment, and labor migration to the United States increased considerably.

With trade liberalization and the dismantling of the long-running public program (CONASUPO) which purchased and distributed Mexican-grown maize, the market for maize and tortillas in Mexico became dominated by a few multinational corporations that prefer to import inexpensive corn from the United States, where the crop is subsidized by the government, rather than purchase corn grown in Mexico. Increasingly, corporations such as Grupo Maseca and Minsa are the providers of pre-made tortillas. This industrialization of production has put smaller tortilla producers out of business and effected a shift in the way tortillas are made. Previously those who made tortillas in small factories (tortillerías) and at home used nixtamalization, whereas the main industrial producers today use corn flour.

Between the start of NAFTA and the tortilla crisis of 2006, when costs soared as a result of the agro-industry's hoarding of corn, price speculation, the rising cost of transportation, and the use of corn to make ethanol in the United States and Canada, the price of tortillas rose an estimated 738 percent (Hernández Navarro 2007). At a meeting between the presidents of Mexico and the United States, which coincided with the tortilla crisis, they rejected the renegotiation of NAFTA but agreed to set up a bilateral working group to study the issue and help smallholder farmers achieve a "smoother transition" when remaining quotas and tariffs on corn imports ended in 2008.

Another change in post-NAFTA Mexico was the emergence of an entrepreneurial sector of corn farmers who took up corn cultivation to gain access to the rural subsidy Procampo. These farmers plant high-yielding varieties on irrigated lands, largely in Sinaloa, and have contributed to an increase in the productivity rate of corn.

While the agricultural and trade policies of Miguel de la Madrid (1982–88), Carlos Salinas de Gortari, Ernesto Zedillo, and Vicente Fox (2000–2006) varied, throughout their administrations a commitment to the neoliberal agenda deepened, affecting corn production and consumption in Mexico. To sum up, the neoliberal corn regime emphasizes food imports rather than national food self-sufficiency and promotes rural development through modern, commercial agriculture, improved seeds, trade liberalization, and the displacement of inefficient campesinos. These policy initiatives, although regionally varied and not always successful, worsened the conditions of many smallholder agriculturalists.

Drawing on earlier images and arguments about peasants and indigenous people, neoliberal government rhetoric bundles together assumptions about corn agriculture, culture, and development. Smallholder corn production is evaluated primarily in relation to market value, and in this narrow frame corn cultivation is seen as an unprofitable, inefficient, and stubborn tradition. It is viewed as archaic and inefficient in Mexico for several reasons: it is generally low-yielding, particularly when compared to its counterpart in the United States; it relies on traditional technology (or a combination of traditional and modern technologies); and it is (or is seen to be) aimed at subsistence rather than exchange or profit. This focus not only diminishes the role of small agriculturalists in feeding Mexico and in the on-farm maintenance of biodiversity, but it overlooks the reasons why rural agriculturalists choose to grow maize.

Kirsten Appendini (1998) has pointed out that the neoliberal policy regime treats smallholder corn production as a drain on scarce government resources, and as needing to "modernize" or "globalize." In other words, corn production is encouraged to become more efficient through modern technology such as scientifically improved seed, tractors, chemical fertilizer, and pesticides or to be replaced with crops that have a comparative advantage for Mexico in the global market, such as fruits and vegetables. In addition to these two options (to modernize or to globalize), the official government position has also sought to encourage rural outmigration and proletarianization. The undersecretary of agriculture, Luis Tellez, said in the early 1990s: "It is the policy of my government to remove half the population from rural Mexico during the next five years." (Barkin 2002, 13).

High-yielding modern and transgenic varieties are also treated as "scale-neutral," or of equal benefit to producers with and without resources. But just as critics of the Green Revolution pointed out, the inputs (irrigation, chemical pesticides, etc.) necessary to produce the high yields of these varieties are often prohibitively expensive for peasants (Barkin and Suárez 1983; Hewitt de Alcántara 1976; Shiva 1991). Access to resources and low-yielding seeds are again reduced to technical problems.

Official rhetoric about corn production reproduces elements of the earlier modernization framework, which argued for incorporating rural peoples into the national project of development through cultural acculturation (which includes economic practices and beliefs). In a newspaper interview the minister of agriculture under Vicente Fox's administration, Javier Usabiaga, suggested that maize-based agriculture is part of a cultural barrier to economic development. He explained that for rural producers today, it is more important to know how to sell agricultural goods than to know how to grow them, that producers need better access to the market, and a change in culture:

> MINISTER USABIAGA: We need to teach the agriculturalist how to sell. They should do things well [las cosas se deben hacer como Dios manda]. We cannot resolve the problem with speeches nor with subsidies. We are going to solve the problem with actions that provide the producer with access to the market. Today it is more important to know how to sell than it is to grow. I want to help overcome the problem of "coyotes" and middlemen, so we will provide financing for trucks.
>
> JOURNALIST: Is it possible to make corn, beans, and coffee viable when they are grown for cultural reasons?
>
> MINISTER USABIAGA: This is a very serious problem that we need to change. We are fighting against a culture. But there are situations in which the grower has been convinced that he cannot continue to do the same thing . . . because even with subsidies it is difficult to earn or cover his costs of production. . . . Today, I have a responsibility to teach them what I learned on my own: to sell, to manage markets, and to make value added products. I'm going to support the entrepreneur. In my view, the entrepreneur is the person who has one hectare [of land], or who has 50 or 1000 hectares. They can't say that I'm going to favor the entrepreneur! Of course, I'm going to do whatever it takes to make everyone an entrepreneur. The person who has two goats, has to know how much they cost him, or how much milk they produce, or what he earns from selling the milk." (La Jornada, 17 January 2001)

Agriculture is of course inseparable from cultural practices and traditions, but Usabiaga views maize-based peasant culture as a particular kind of culture, one which is a barrier to economic development because it overvalues subsistence (rather than commercial production or profit making). The minister suggests that peasants need to become entrepreneurs. Maize agriculture is also culture in the sense that it is a stubborn and problematic (as in inefficient or backward) tradition, suggested by the statement that the government "is fighting against a culture" and that the culture is "a very serious problem." The minister indicates that many rural producers do not adapt well to new values and technologies, although some have: "There are situations in which the grower has been convinced that he cannot continue to do the same thing."

The minister's concern with teaching agriculturalists how to sell directly to "the market" is partly due to the need to bypass the exploitative intermediary. Intermediaries pay unfairly low prices for agricultural produce to the rural producers who have little means to transport their goods to urban consumers. In the Tehuacán Valley, for instance, farmers rely on intermediaries to purchase and transport their corn. In the official framing of the issue, as suggested in the minister's quote, the problem of rural poverty is a problem of isolation from the market, rather than the result of how rural regions have been shaped through wider processes of capital accumulation, or more recently, how deregulated and unsubsidized access differently affects producers of disparate resources and locations.

The minister's perspective is reminiscent of earlier agrarian debates in which the "natural economy" of peasants was seen as largely isolated, millennial, and self-provisioning. In this view peasants and rural indigenous peoples stubbornly undertake subsistence production even when alternatives are available because they distrust outsiders and the accumulation of wealth within their communities. While these practices were seen by some as the result of isolation from external markets, modern cultural influences, and state policies, other anthropologists in the agrarian debates argued that community leveling and corporatist organization were the result of historical processes. While specific peasantries may indeed be isolated and subsistence-based, peasantries are always historically and socially produced (Roseberry 1989).

Also connected to the layered assumptions about corn-producing peasants is the way rural "development" is conceptualized. Rather than view rural poverty as the historical product of unequal economic and political relations, official policy views rural Mexico as lacking sufficient access to the market. Development is thus seen as providing that access and as transferring technology from the global north to the global south so as to increase levels of production, consumption, and economic growth.

The term "development" often refers to both a process and a goal. In *Doctrines of Development* (1996) Michael Cowen and Robert Shenton locate the historical roots of the modern idea of development in the social upheavals of the first half of the nineteenth century in western Europe, during the Age of Revolution. Projects for development emerged as palliative efforts (or revolutionary ones, to Saint-Simonians and other socialists) to remedy the process of capitalist development, particularly the creation and re-creation of surplus labor. Cowen and Shenton contend that in both early and contemporary theories of development, "The welfare of the relative surplus population became the aspirant of development" (1996, 475). And *the intent* of dominant doctrines of development has been conflated with *the actual processes* of capitalist development. As a result, these doctrines seek to ameliorate poverty and social disharmony through economic growth, increased levels of production, and expanded opportunities for waged employment.

In a similar fashion, the neoliberal approach to rural Mexico, along with its modernizationist predecessors, locates the origins of rural poverty in limited access to markets and inefficient production while overlooking the ongoing creation of rural surplus labor in the expansion of capitalism. A key difference between earlier rural policies and the neoliberal corn regime, however, is that the latter advocates the free hand of the market rather than state intervention in distributing resources.

Rather than looking to views of corn culture (as a stubborn set of traditions and values which are inefficient) for explanations about why corn agriculture matters to rural producers, during my fieldwork I came to see corn production as part of a strategy of the relative surplus population to cope with accumulation by dispossession. Maize is, as discussed in later chapters, part of the livelihood strategies of valley households as they contend with crisis.

Some members of In Defense of Maize go so far as to suggest that smallholder criollo corn agriculture is a form of resistance to neoliberalism, but I would caution that while corn production and biodiversity are a political issue and part of a politicized identity, and even a form of resistance for some campesinos and indigenous groups, this is not true of all rural Mexicans. Farmers grow corn for a multiplicity of reasons, some formulated and articulated, others not. Farmers may grow corn in general and criollo corn in particular because it provides a needed source of food and possible income for a generation of older residents who are no longer employable or cannot easily migrate north. (This is suggested by Sanjosepeño narratives and practices explored in chapter 5.) They may also grow corn because it is a tradition or a taken-for-granted part of rural life.

As for the relative merits of criollo and scientifically improved maize, smallholder producers may prefer a variety of seed which does not have to be purchased at every planting cycle so as to maintain its yield, as is generally true of modern varieties. Criollo seed typically has a lower yield than scientifically improved varieties, but criollos can be saved and replanted by farmers at no expense and with no detrimental effect on yield (Soleri, Cleveland, Aragón, Fuentes, Ríos, and Sweeney 2005).

Second, small producers prefer varieties which perform well in the particular climate and soil conditions of their own area. Criollos have adapted over time to specific and sometimes difficult agro-environmental conditions. Producers may prefer local seed because they trust information about its performance from known sources, or because the seed has been grown nearby in a similar environment (Perales, Benz, and Brush 2005, 953). For these reasons criollo maize is a peasant crop par excellence: it is more stable, tried, and trusted seed, which maintains a degree of farmer autonomy or control over production, lessening the interference of agribusiness and the state.

A third possible consideration is that while some criollos may require chemical pesticides and fertilizers to produce a good yield, this requirement is inescapable for improved seed, and small-scale farmers may not always have the funds to buy the needed inputs. And fourth, farmers may prefer the taste or texture of criollo maize over improved varieties, as they do in the southern Tehuacán Valley. In conclusion, to focus solely on high yields and market access, and to regard culture as an inefficient barrier to development, misses the reasons why criollo production might be practiced, and even preferred, by rural Mexicans.

THE CALL FOR MAIZE SOVEREIGNTY AND QUALITY

At the first forum held by In Defense of Maize in Mexico City in January 2002, participants argued that GM corn contamination must be situated and understood as part of a larger attack on peasant livelihoods. This multidimensional approach asserts that peasants are maize experts, as we saw in chapter 1, and challenges the neoliberal (and before it, the high modernist) model, which sees smallholder corn agriculture as inefficient and backward. The comments at the forum of a representative from the Support Center for Indigenous Missions reflect this approach: "With Progresa [the anti-poverty program] and Procampo [the transitional agricultural subsidy], with all the agrarian and food security programs there are in Mexico, it is estimated that the economically active population in the countryside will be reduced from 27 percent to 3 percent in less than ten years. That is to say, [the government's perspective is:] we don't need peasants, nor do we need indigenous communities. We need people that can work in the *maquiladoras*. This is the solution that the neoliberal government wants to propose to us" (tape recorded, Mexico City, 24 January 2002). To counter the neoliberal agenda, In Defense of Maize has forged alliances with other peasant, environmental, and indigenous organizations and campaigns, such as the huge campesino demonstrations against NAFTA in late 2002, which took place under the banner "El Campo No Aguanta Más.[4] The multidimensional approach of the coalition to the issue of GM corn is reflected in the conclusions and demands of the first and second In Defense of Maize forums. They address not only contamination and the need to protect traditional maize varieties but broader agrarian policies. Participants at the second forum, for instance, suggested that migrants be included in discussions of the problems facing rural Mexicans and the use of remittances.

En Defensa del Maíz Forum, January 23 & 24, 2002, Mexico City
Demands to the Mexican government include:
—To declare maize as a strategic resource of national security. To establish policies that protect and encourage its cultivation since maize is the basic food for Mexicans and central to their symbolic and material culture.
—To reconsider agrarian and commercial policies. In particular, to review NAFTA's chapter on agriculture and on basic grains in order to reverse the effects of systemic dumping on Mexican peasants' agriculture.
—To suspend the import of GM corn from the United States immediately. Such imports have been identified as the main source of contamination of Mexican maize. The *de facto* moratorium over the com-

mercial release or experimental cultivation of GM corn should become legally compulsory.

—To expel the multinational corporations responsible for the spread of GM corn, such as Monsanto, Novartis, DuPont and Aventis from the country and demand that they carry the costs of decontaminating the affected areas and compensating the affected population.

—The government should finance the recovery and use of traditional seeds within local communities.

—To recognize the principle of food sovereignty based on the strength of the peasant economy and the rights of the indigenous and agriculturalists.[5]

En Defensa del Maíz, Second Forum, December 2003
Conclusions include:

—Promote the defense, recognition and spread of traditional techniques of cultivation (agronomic, ecological, medicinal and so on) and new techniques of organic production as part of the long term process of decontamination.

—Promote the cultivation of known seeds.

—Strengthen indigenous and peasants' autonomy and community organization; defend maize alongside the struggles for their territory and self-government.

—Maintain the moratorium on cultivating GM corn at the national level.

—Boycott food aid packages and soft drinks which contain GM corn or GM corn syrup.

—Promote food alternatives for rural and urban populations and consumers organizing.

—Reject the law on biosafety and demand the inspection of imported grains.

—Engage with migrant organizations in order to dialogue on shared problems as well as the importance of maize and remittances.

—At the international level, demand and promote local and native corn varieties as the heritage of humanity; and frame the struggle over maize as a human rights struggle.[6]

The coalition also contributes to and draws from transnational environmental, peasant, and food sovereignty networks and movements. In the 1990s an international network of activists and NGOs began to articulate a link between agricultural biotechnology, peasant, labor, and environmental issues as part of its critique of corporate-led and neoliberal globalization. Heller (2002, 14–15) succinctly explains the rise of this perspective:

> Since the 1980s, international environmental NGOs associated with such publications as the Indian based magazine, *Third World Resurgence* and the British journal, *The Ecologist*, had been developing a cultural and political economy critique of GMOS. Between the 1992 Earth Summit in Rio de Janeiro, and the 1999 anti-WTO demonstrations in Seattle, these cultural and political economy perspectives on North/South relations, international peasant movements, and "sustainable development" expanded to become what could be called an "anti-globalization critique" associated with the international anti-globalization network. Within this network, questions surrounding GMOS, such as agricultural biotechnology, biodiversity, and intellectual property, are located within a broader rubric of global capital, global peasant movements, trade liberalization, and international environmental policy.

Alter-globalization networks and organizations have highlighted how the fate of rural peoples is tied to the global food regime. The transnational peasant organization Via Campesina, formed in the early 1990s, has argued that the "massive movement of food around the world is forcing an increased movement of people."[7] This is certainly true in Mexico, where the erosion of food sovereignty has contributed to an accelerated migration from the countryside to urban Mexico and across the border. Peasant and food sovereignty organizations and movements draw attention to people marginalized by the expansion of agribusiness and the privatization of state services; they are, in other words, a political response to accumulation by dispossession (McMichael 2006).

The Via Campesina, one of the participants in the Mexican anti-GM corn coalition, is composed of national and regional agrarian organizations which have joined forces across national borders. The organization denounces the promotion of food security through imports (Desmarais 2002). Instead of "food security" the Via Campesina promotes "food sovereignty," which it defines as "the right of each nation to maintain and develop its own capacity to produce its basic foods respecting cultural and productive diversity. We have the right to produce food in our own territory. Food sovereignty is a precondition of genuine food security" (cited in Desmarais 2002, 104).

This concept of food sovereignty challenges the conventional model for agriculture, neoliberal policies, and the global food regime. João Pedro Stédile (2002, 100), a regional coordinator for Via Campesina in Brazil, explains that the concept of food sovereignty "brings us into head-on collision with international capital, which wants free markets. We maintain that every people, no matter how small, has the right to produce their own food. Agricultural trade should be subordinated to this greater right. Only the surplus should be traded, and that only bilaterally. We are against the WTO and against the monopolization of world agricultural trade by the multinational corporations. As José Martí would say: a people that cannot produce its own food are slaves; they don't have the slightest freedom. If a society doesn't produce what it eats, it will always be dependent on someone else." Transnational peasant and food sovereignty movements forge alternatives to conventional agriculture by making alliances between food producers, distributors, and consumer and environmental groups. These "alternative geographies of food production" promote an economy of quality over one of scale and efficiency (Whatmore and Thorne 1997). They value "food from somewhere" or place-based foods, grown in season, in particular environments, and based on particular (artisan or métis) expertise; this is quite distinct from the corporate or neoliberal food regime's "food from nowhere" (McMichael 2009, 147).

The slow food movement, initially founded in Italy in 1986 and now transnational, also advocates for food quality. It promotes "good, clean and fair food," based on a belief that the food we eat should not only taste good but be produced in a way "which does not harm the environment, animal welfare or health; and that food producers should receive fair compensation for their work."[8] In other words, there are three dimensions to food quality: the taste or quality of the food, a reduction of harm to the natural environment, which includes the protection of traditional varieties from agribusiness and industrial agriculture, and the quality of producers' livelihoods.

This multilayered notion of food quality informs the support of In Defense of Maize for alternative food networks. The networks aim to support small farmers who plant criollos while also providing consumers with good-tasting, regionally grown, non-GM-corn-based foods. In Oaxaca one tortillería and restaurant has served regional delicacies made from criollos to customers since 2001. In the process the owners have sought to promote "a deeper appreciation of biodiversity and traditions" in Mexico.[9] GEA, Greenpeace Mexico, and several other NGOs have organized non-GM food trade fairs and launched a logo to indicate when a product is GMO-free (interview with Catherine Marielle, Mexico City, 27 July 2006).

There are tensions and contradictions, of course, within alternative food networks. The differences in priorities and perspectives among producers, distributors, consumers, and environmental groups in various regions and countries are not easily overcome. In the Tehuacán Valley, as in many places, the interest of consumers and producers differ: food shoppers hope for lower prices while producers seek higher prices for their grain. Similarly, class differences among producers can also generate tensions. Small-scale maize producers and northern, large-scale farmers in Mexico often have conflicting perspectives on agricultural and trade policies. Or, to provide another example, local "slow food" chapters operate in some places more like gourmet or "foodie" clubs than associations seeking to enhance the quality of food producers' lives.

Maize and the Multicultural Nation

While corn agriculture has reemerged as a target of state modernizing efforts, a growing number of restaurants and urban consumers celebrate maize-based foods, long associated with Indians, as part of a multicultural Mexico. This "authentic Mexican food explosion" in places like Mexico City involves the production and marketing of organic and criollo-based tortillas and foods (Vizcarra Bordi 2006). In contemporary urban Mexico there has been a rediscovery and celebration of cultural differences and culinary traditions, in contrast to the post-revolution and postwar celebration of cultural homogeneity through racial and cultural hybridity. Among critics of GM corn the language of cultural diversity, plurality, and multiculturalism is employed to emphasize cultural interaction and the coexistence of different cultures over indigenous acculturation, the focus of earlier ideas about hybridity. Both maize foods and maize agriculture are a symbol of the cultural distinctiveness of peasant and rural indigenous Mexico.

Cultural diversity is seen as generating and maintaining the biological diversity of maize. Gustavo Esteva, one of the editors of the anti-GM, pro-maize book *Sin maíz, no hay país*, suggests that maize is a key metaphor for rethinking the project of neoliberal development. Maize can be seen as a "hospitable" crop because in the traditional Mesoamerican milpa, beans and squash are grown among fields of corn. As such a hospitable crop, maize is a metaphor for *convivencia*, or what Esteva refers to as a world respectful of different ways of life. As a plant with so many types and varieties, maize is also a metaphor for not sacrificing diversity to the homogenizing forces of the global market or the homogenizing forces of industrial agriculture, but finding creative, multiple answers in diversity. "In culture, as in nature, diversity embodies the potential for innovation and opens the path to creative solutions" (Esteva 2003, 300; Engl. trans.). The celebration of cultural diversity as a critique of conventional agriculture is not restricted to Mexico, but is informed by transnational environmental and alter-globalization movements. Vandana Shiva, in her well-known criticism of the Green Revolution in India and GM crops, has similarly written, "Cultural diversity and biological diversity go hand in hand." (1993, 65).

In Defense of Maize challenges the neoliberal model by reframing the issues and re-politicizing questions that have been rendered technical. But it also can slip into peasant essentialism and a bounded, reified conception of culture. In *Sin maíz, no hay país* Esteva employs the notion of "deep Mexico"—a term coined by the anthropologist Guillermo Bonfil Batalla (1996 [1987]) to refer to millennial indigenous culture—to talk about modern Mexico as "the coexistence of two distinct civilizations: the Mesoamerican and the occidental" (2003, 297). To Esteva the worldview of Mesoamerican civilization is based on the values of community and subsistence, while the occidental worldview is based on the market, profit, and exchange. In this Esteva shares the idea of corn culture, or deep Mexico, with the neoliberal model he opposes. The official, neoliberal narrative posits peasant and indigenous corn production as inefficient in part because it is deemed a culture of subsistence unfamiliar with the entrepreneurial spirit and the profit motive, while critics counter that the values of subsistence culture are a beneficial alternative to the market and its processes of commodification. The tendency of these critics to see corn culture as a natural economy opposed to capitalism romanticizes it as egalitarian, in harmony with nature, and resistant to outside influences forces (O'Brien and Roseberry 1991, 3). This opposition can slip into a notion of primary and original cultural alterity and overlook how indigeneity and peasantries are made and remade in interaction with larger forces and processes.

Another characteristic of this essentialism can be found in the portrayal of peasant and indigenous agriculture as eco-friendly. At a press conference held in Mexico City in March 2001 by the Via Campesina, Mexican representatives welcomed the French farmer and activist José Bové to Mexico. At the press conference Bové, who is also well known for protest actions in France at which transgenic maize has been destroyed, explained his visit by saying that because of the movement for a GE-free agriculture and the uprising by the EZLN (Ejército Zapatista de Liberación Nacional) against NAFTA, Mexico was at the center of alter-globalization struggles. When a representative from La Via Campesina spoke, he explained that Mexican indigenous producers and peasants "have always existed in harmony with nature" (11 March 2001).

Statements about living in harmony with nature are often made to highlight the dependency of peasants and rural Indians on their environment or to foreground their uncompensated contributions and knowledge to crop biodiversity and in situ conservation. Such portrayals of peasants and rural Indians can be a political gesture to emphasize their unpaid and uncompensated labor in the adaptation, selection, and modification of criollos and the related production of potential economic value. Although the Via Campesina promotes a holistic approach to food quality and agricultural livelihoods, and although it is an organization representing a multiplicity of opinions and members from around the world (Desmarais 2002), at times corn-based livelihoods are presented as a millennial culture of subsistence, always protective of the natural environment.

Among activists, academics, scientists, industry, and officials in the GM debates, corn culture—values, practices, agricultural knowledges—is often treated as fundamentally distinct from urban, market, and modern cultures. The struggles over and narratives about the fate of the countryside, and indeed the nation, frequently draw upon anthropological constructions of culture, as suggested above. Within anthropology there have been many criticisms of the classic "billiard ball" model of culture as an autonomous entity with an internally coherent system, pattern, or logic. In recent years anthropologists have tended to highlight the unboundedness of culture and communities, particularly in a globalizing world. Despite these efforts, a bounded view of culture manages to creep back into representations of rural Mexico, obscuring the ways peasant labor and peasant agriculture interact with larger processes. One of the tasks for anthropologists is to continue the critique of our own traditions in conceptualizing culture, while also investigating how the concept of culture is used in making political claims.

Corn-based foods like tortillas have become one of the most powerful symbols of what it means to be Mexican, and all the contradictions that being Mexican entails. While at mid-twentieth century corn-based foods went from being a symbol of indigenous backwardness and isolation to a symbol of the mestizo nation, in state development projects and anthropological debates corn agriculture continued to represent indigenous backwardness or the natural economy of peasants. In more recent years GM critics have used maize agriculture as a symbol of cultural distinctiveness in the hybrid nation: as part of a multicultural hybridity rather than a culturally or ethnically homogeneous nation. At times, however, coalition narratives, like those of the Ministry of Agriculture, include older frameworks about peasants and indigenous peoples as part of a natural economy or millennial cultural tradition. In other words, both proponents and opponents of transgenic corn rely on a notion of corn culture and peasant essentialism.

Despite any shortcomings, In Defense of Maize challenges official narratives about inefficiency and the logic of import dependency. In contrast to a neoliberal model which seeks to transform campesinos into entrepreneurs, promotes food security through imports, and emphasizes efficiency and the quantity of corn produced, the anti-GM coalition draws upon transnational movements and the Mexican counter-tradition of valuing peasant knowledge to advocate for maize sovereignty and quality in terms of its taste, its role in in situ biodiversity, and the quality of the producer's livelihood. Maize is more than an economic good: it is a multidimensional crop and food, with multiple meanings and purposes. This will be made clear as we examine producers' livelihoods in the southern Tehuacán Valley, one of Mexico's "cradles of corn."

Part II LIVELIHOODS

In the southern Tehuacán Valley the cash crop production of fresh corn (or elote) took off in the 1960s and 1970s and became popular among producers in San José Miahuatlán a decade later. Although cash crop production is widespread in the region, maize is clearly more than an economic good. Various parts of the corn plant are used to decorate the entrances to homes during religious celebrations, signaling the use of maize in important rituals and its former spiritual significance. The name "Miahuatlán" means "the place of the flowering corn spike," or ear of corn. It has been interpreted to mean the place of Centeotl, the Aztec goddess of the land and agriculture, because she is represented by this part of the plant (Paredes Colín 1921, 66). Maize production and preparation also constitute part of the taken-for-granted, gendered work of the household which structures everyday life. The male domain of the cornfield (milpa) and the female domain of tortilla preparation help to demarcate the division of labor in the household and the community along gendered and generational lines. Farmers just starting out inherit criollo seed selected from their parents' or older relatives' fields.

Maize agriculture is central to subsistence strategies, not as a manifestation of a static natural economy or deep Mexico but as an adaptive tradition and flexible form of cultivation. The production of maize is a "strategy for maximizing household labor" (González 2001, 164; Warman 1980 [1976]) and contending with the process of "accumulation by dispossession." The cultivation, preparation, and consumption of maize are also "cultural" practices. In the valley residents engage in the practices and concepts they have inherited, carrying some elements forward, leaving others behind, and incorporating new experiences and ideas along the way. A changing natural environment shapes possibilities and places constraints on social organization and agricultural practice.

San José is a transnational town: everyday life is affected by the experience of labor migration. Many Sanjosepeños work in the United States yet remain engaged in the lives of friends and family back home, and contribute financially to their valley households. At the same time, residents' ideas about being a Sanjosepeño are rooted in the history and practices of the town. Place-based practices shape the production and reproduction of culture. To understand how neoliberal policies and globalization have affected maize production and livelihood practices in San José—the topic of chapters 4 and 5— it is necessary to have a sense of "what has been historically sedimented there, on the social and spatial structures that are already in place" (Pred and Watts 1992, 11). For this reason chapter 3 focuses on the social terrain of community in San José, examined through various scales of analysis, as a product of interaction with regional and state as well as transnational processes and projects. Communities like San José are not just the products of transnational processes: they also determine how transnational processes and projects materialize. Maize production and irrigation management have contributed to the organization of community as a political administrative unit, a set of social institutions and unequal relations, and an imagined sense of shared belonging.

In other regions of Mexico maize can still have strong spiritual and ritual significance or may be connected to a positive and politicized indigenous identity. In the Sierra Juárez of Oaxaca, for example, Zapotecos view maize as alive, as having a soul, and ascribe human qualities to the plant (González 2001). In contrast to the Tehuacán area, when GM corn was found in Zapotec highland communities the contamination became a political issue, in part because maize is central to Zapotec indigenous identity. In recent years this indigenous identity has been reclaimed and linked to other Zapotec communities as well as to a pan-indigenous political movement (Kearney 1996; Stephen 2005). As we shall see, this positive and politicized indigenous identity is distinct from how Sanjosepeños describe themselves and discuss indigenous traits.

In the southern valley the news that evidence of transgenes had been found in the Tehuacán area went largely unnoticed. Sanjosepeños were unfamiliar with the controversy which made international news and inspired protests, public forums, and academic conferences. Not one maize farmer whom I interviewed in 2001–6 had ever heard of *maíz transgénico*, nor was it a topic familiar to the regional employees of the Ministry of Agriculture with whom I spoke. As residents are the de facto guardians of the in situ biological diversity of criollos, their perspectives and practices are relevant to the debates about GM corn and the future of maize. Although they do not discuss their agricultural production as being the work of guardians, residents are concerned about their livelihoods and the future of agriculture in the valley.

When Sanjosepeños discuss their livelihoods and maize agriculture, they do so in terms of a different set of risks and challenges. They mention the challenges posed by recent cuts to rural subsidies, the lack of employment opportunities, agricultural labor shortages, declining maize yields, and decreasing spring water levels. Residents of all ages value local criollo corn for the taste and texture of tortillas made with it, but worry about who will produce criollo corn in the future. The cultivation and preparation of maize as food help to link households in social relations of obligation, reciprocity, and exchange. There is however a growing generational tension in perspectives about maize varieties and production. Older residents continue to value local maize for its taste, and for providing a kind of social safety net, while younger residents tend to view the crop as unprofitable. Chapters 4 and 5 show how maize agriculture is changing as Sanjosepeño households adapt to declining water rates and the neoliberal corn regime.

Chapter 3

COMMUNITY
AND CONFLICT

This chapter explores the formation of San José Miahuatlán as a community with lines of solidarity and tension. San José is the head town (*cabecera*) of the county (*municipio*) of the same name. Each state in the republic is divided into political-administrative units called municipios, which are governed by a mayor (or *presidente municipal*) and his municipal council (*ayuntamiento*). But San José is also a community in ways above and beyond its status as a political and administrative unit or geographic locale; it is organized, experienced, and imagined in particular ways. Residents participate in kin-based social and economic institutions which emerged in the interaction between indigenous people and colonial powers, such as *compadrazgo* (ritual kinship), the *mayordomía* (a civil-religious cargo system dedicated to honoring saints), and the *faena* (communal labor). These institutions are interlinked with maize cultivation and preparation and have implications for social arrangements *within* the household along generational and gendered lines (in the milpa, domestic duties, and service to the community) and relations *between* households (in ritual kinship, the civil-religious system, sharecropping, and water management).

These social and economic, largely kin-based, institutions bind individuals and households in several overlapping ways: they structure the rights and obligations of residents, access to water and land, and notions and practices of gender, age, and ethnicity, as well as providing avenues for gaining respect and social prestige. In other words, they help structure the social relations of production and reproduction, and various axes of social difference. In her discussion of these institutions among the Zapotec of Teotitlán, Oaxaca, forty-eight kilometers south of Tehuacán, Lynn Stephen draws on the concept of social reproduction to examine how gender intersects with relations of class, age, and ethnicity. She explains how the concept of social reproduction, adapted from Christine Gailey's work, helps to "focus not only on the maintenance of material relations of production but also on the maintenance and replication of institutions and relationships that define individuals as social actors in their specific ethnic context" (2005, 57). Stephen goes on to explain these institutions in relation to social reproduction (ibid.): "These kin-based institutions are the backbone of ethnicity; they are also crucial in the reproduction of the labor relations of class. In this sense, the concept of social reproduction specifically includes reproduction and socialization of the labor force, including relationships linked to biological kinship, compadrazgo or ritual kinship . . . and institutions, events, and relationships tied to the socialization and status of individuals, primarily through mayordomías, life cycle ceremonies, and the reciprocal exchanges of goods and labor that support them." Similarly, these institutions of the mayordomía, compadrazgo, and faena are the backbone of ethnicity among Sanjosepeños. Residents have a sense of shared ethnicity with other valley residents based on these institutions, language use, and social and economic marginalization from the wider Mexican society, while also identifying as a particular "people" (pueblo) in terms of dialect, traditional dress, and marriage rituals. Sanjosepeños speak a dialect of Nahuatl, which differs from the dialects spoken in other parts of the valley. Most residents today also speak Spanish.

For many years the official markers of Indian status were language and residency in an indigenous village. Residents who left their community and spoke Spanish were expected to assimilate and become mestizo (Martínez Novo 2006). Today Indians remain popularly associated with the countryside, and with poverty and social immobility.

Everyday understandings of ethnicity in Mexico, as in Latin America more generally, are often informed by a discourse of race in which cultural or ethnic differences are essentialized (Poole 1994, 2, 6). In the valley the category of Indian not only is informed by essentialized differences but can overlap and be strategically contrasted with a campesino identity forged during the revolution. Christopher

Boyer suggests that the campesino identity appealed to many disenfranchised rural Mexicans of the early twentieth century—and continues to resonate in the early twenty-first century—because it reflects a collective memory of oppression under the hacienda system, and because it does not require casting off other deeply felt identities (2003, 235). As a broad, inclusive, and political identity, campesino is a category amenable to multiple and composite cultural attachments. Like the category of Indian, campesino can be both ascriptive and a deeply felt identity. In San José residents refer to campesinos when highlighting the quality of being hardworking, or *indios* and *indígenas* when highlighting a tendency toward violence, but both identities imply being unjustly disadvantaged economically and politically.

Sanjosepeños regard themselves as sharing with their valley neighbors certain cultural traits and a predicament of poverty, injustice, and discrimination. At the same time, they set themselves apart from their neighbors—and are set apart by others—as *more* Indian, in the sense that they live in a poorer, more desperate, and more violent town. So although the category of Indian connotes a degree of shared experience, it is not shared in the same way by different individuals or even in the same ways by the same individual over time (Stephen 2005, 18). In the valley the sense of a shared indigenous ethnicity is clearly not uniform across the region but is circumscribed by the history of each town and the class relations within and across towns, and by notions of hierarchy and contradiction (Macip 2005).

Sanjosepeños tend to refer to themselves as campesinos rather than using the label "indio" or the more neutral "indígena." Yet at the same time, residents view the history of local conflict over water in relation to an essentialized, even racialized ethnicity, which mobilizes older prejudices about the violent nature of Indians. Rather than interpret residents' use of the language of violent Indians as simply the internalization of external racist discourse, I argue that the representation of Indians as violent is sometimes mobilized by residents in defense of their resources, while at other times they employ a campesino identity. Community and ethnicity are explored here in relation to particular political and economic contexts (Martínez Novo 2006; Macip 2005; Poole 1997; Roseberry 1996; Stephen 2005). More specifically, to understand what it means to be a Sanjosepeño, we need to look at the struggle over resources in the region.

THE BARRIO SYSTEM
OF WATER MANAGEMENT

In San José the mayordomía, compadrazgo, and faena are connected to a barrio or neighborhood system of spring water distribution and management. This older system of distributing natural spring water (manantiales) by neighborhood was the principal means of distribution in San José until the mid-twentieth century, when valley residents increasingly invested in the construction of underground chain wells (galerías filtrantes) and deep wells which rely on electric pumps.

The development of this highly organized and complex irrigation system of galerías and wells was relatively independent from state control, expertise, or financial support; it is a largely self-managed and decentralized system, with members of communal associations (sociedades explotadoras de agua) managing the galerías and canals (Henao 1980; Enge and Whiteford 1989).

Although surface and groundwater became state property after the revolution, individuals and groups that constructed and maintained irrigation facilities, like the valley water associations, could be granted water use rights (Aboites Aguilar et al. eds. 2000). Elsewhere irrigation management was centralized during agrarian reform, but in the valley irrigation was largely self-managed, and the state rarely intervened directly. This was not only because the system was managed efficiently and the state granted use rights to people who constructed and maintained irrigation facilities, but because of strong opposition to the nationalization of water in the valley (Enge and Whiteford 1989, 184). To say that valley irrigation has been a largely self-managed system, though, is not to downplay the effects of state policies and practices in the region. In a pivotal decision for San José, the post-revolutionary state denied residents' petition to reclaim two important natural springs, Atzompa and Coyoatl, in 1936.

The construction of galerías and deep wells by communal water associations and businesses rapidly increased between the mid-1940s and the late 1960s.[1] Although galería associations were a regional adaptation to the increasing competition for diminishing water resources, their proliferation, along with that of agribusiness, has contributed to the overuse of water in the valley. Tehuacán's famous spring-water bottling companies rely on galerías. The capital-intensive aviculture industry, established in the valley by mestizos from Tehuacán in the 1950s, also depends on water from springs and pumping stations. And in the 1990s the region experienced a boom in the installation of maquiladoras, which required large quantities of water for their operations. As a result of all these demands for water, the rate of depletion of the water table has increased. The region's natural springs and some galerías are affected, particularly in the valley's southern end.[2] Galerías, not natural springs, are now the main source of irrigation water in the valley.[3]

The drying up of one of San José's principal sources of water, the Atzompa spring, in 1954 profoundly affected agriculture and community relations, including the mayordomía. After the drought a group of residents formed an association to build a galería filtrante, but when water was found in the early 1980s it prompted violent conflict between the water association and the municipal government.

In this chapter we consider how the decline in water flow rates contributed to a tension between the older system of barrio management in San José, which relied on spring water, and the newer system of associations which tapped the water table through galerías and deep wells. This was a tension between the claim to groundwater as a state-owned or public resource and the claim to private access to groundwater based on investment.

Water associations were a way for residents to regain access to water and tap into the water table as the flow at springs declined, but they also contributed to the problem of lower water levels. This was a "tragedy of the commons," but not one that arose because individual residents acting in their own self-interest overexploited a communal resource, as suggested in the polemical essay by Garrett Hardin (1968), but rather one that arose because large landowners and later industry were empowered to overexploit water resources and did so. It was in this context of appropriated and overexploited water resources that valley residents responded in creative and sometimes violent ways. In San José declining water levels and the accompanying conflict affected agricultural production and also the town's main social and economic institutions. In turn, the conflict drew upon and reinforced essentialized generalizations about Indians.

Historical Background

Many of the valley's towns, including Miahuatlán, existed in some form in pre-Hispanic times. In the early sixteenth century the valley came under the purview of the independent state of Teotitlán del Camino, allied to the Aztec empire (Byers 1967). At the time of Spanish conquest there were settled populations in the valley that had developed an irrigation system using the area's natural springs and the Río Salado (the name given to the Tehuacán River because of its high salt content). Sections of an ancient canal system, along with other canals built before conquest, were used until the seventeenth and eighteenth centuries (Henao 1980, 61). Built between 150 B.C.E and 700 A.C.E. during the late Palo Blanco phase, some sections of the ancient canals remain in the valley today, preserved by calcium carbonate deposits from the water that ran through their inner walls. In Nahuatl these canals are called "stone snakes" (tecoatl), reflecting the fact that they wind their way down the valley floor (Woodbury and Neely 1972). These canals may have formed one long system that used the valley's natural springs as its water source (Woodbury and Neely 1972). Many of today's irrigation canals run parallel to these older systems (Gil Huerta 1972).

The Spanish introduced the production of wheat and sugarcane to the valley in the sixteenth century, using land and water confiscated and usurped from Indian communities. In an effort to facilitate colonial administration and the extraction of labor and tribute, dispersed indigenous populations were obligated to relocate to existing valley communities, where they were allotted limited land. As part of this process different ethnic groups (Nahuas, Popolocas, Mazatecos, and Mixtecos) were placed together in towns organized by neighborhood. The Spanish also brought in indigenous leaders (caciques) from the state of Tlaxcala to help administer the valley and oversee Indian labor. These caciques received land and were exempt from taxes. Over the years they amassed considerable wealth and prestige (Henao 1980, 62; Paredes Colín 1921, 96). Indian residents of valley towns did not own spring water but had access to the resource through their barrios. The division of towns into different, self-contained barrios helped facilitate political control of the indigenous population and the distribution of water. It also generated conflict over resources within the communities (Henao 1980, 32–57).

Like other valley towns, the cabecera of San José was divided into four main neighborhoods. Residents believe that these barrios were settled at different historical moments:[4] first Miahuatla, followed by Ateopa (or Ateopan), Atzacoalco, and then the largest barrio, Tuxpango. Today each barrio is responsible for the care of two religious images (as part of the mayordomía), runs a committee for neighborhood affairs (mesa directiva), and has access to an irrigation canal. An exception to this, at least during my fieldwork, was the new barrio Guadalupe, founded in the 1990s on formerly agricultural land with migrant remittances.

The four neighborhoods of San José were more than just physical places for residences, functioning almost as completely self-sufficient communities. In addition to being territorial, each barrio was endogamous and organized according to patrilineal descent and patrilocality. Older residents recount how men and women did not marry across barrio lines until the mid-twentieth century. Today, when a woman marries a man from a different barrio she becomes a member of his barrio and may move into his parents' house, as was done in earlier times. According to an emergent trend, young wives live in the homes that they establish with their husbands, built with the couple's earnings as migrants and maquiladora workers, rather than with their parents-in-law.

Sanjosepeño households still rely on male marriage brokers (tetlallís) in their interactions with each other and with the community authorities. Tetlallís, who are considered articulate and wise, represent the groom in asking the bride's parents for permission to marry. Previously the tetlallí was also a respected go-between for families and represented heads of household in important interactions with civil and church authorities. Some tetlallís spoke Spanish, a particularly useful skill in what was a largely monolingual town. This role as broker or representative was important in a municipio where few residents had completed grade school. The first school was built in the 1930s. These days approximately 80 percent of the population has left school before the age of fifteen.[5] Marriage brokers were once common throughout the valley, but in other municipios with earlier and higher rates of bilingualism than San José the importance of the broker's role has declined.

In colonial Mexico "Indian" was a racial and legal category; Indians were considered the descendents of original inhabitants and were legally required to provide labor and tribute to the Spaniards (Lomnitz 2001, 41). Income and labor were produced for the Crown in two ways: firstly through the encomienda system by which communities had to pay tax and tribute in exchange for the "protection" of the Crown and their conversion to Christianity. Secondly, after the large encomiendas disappeared, the *repartimiento* system conscripted indigenous labor to work the nascent haciendas of the region, the first of which was established in the valley in 1622. Haciendas often faced shortages of labor, which were remedied through decrees that obligated communities to supply a quota of workers to the haciendas, or forced Indians to work or face prison (Enge and Whiteford 1989, 80).

The mayordomía was introduced by the Catholic church to extract money more efficiently from indigenous populations and to foster support for the church among them. By the late nineteenth century and the early twentieth, however, the mayordomía had come to "acquire the characteristics of self-identification and indigenous opposition" to mestizos (Henao 1980, 84; translated from Spanish). This civil and religious cargo system, still in use today, involves a hierarchy of rotating posts and the assignment of a steward (and helpers) to care for one of the saints of his neighborhood, who is represented by a religious image. Each barrio has two saints and two mayordomos. These cargo holders contribute to the running of civil and religious affairs in the community. Those who become mayordomo incur the cost of caring for the image, including the cost of important saint's day celebrations.

The colonial tribute and tax system undermined the self-sufficiency of valley Indians and generated disparity within the indigenous communities. There was a growing division of those who accumulated resources—an upper peasantry, which included the caciques from Tlaxcala—and those who relied on subsistence production or paid agricultural labor. Either through obligation or need, a stratum of Indians became dependent on sharecropping and work as agricultural laborers (*peones* or *mozos*), such as seasonal cane cutters. While some Indians remained linked to the community and continued to receive water and land through participation in institutions like the faena, others had their access to community resources and institutions cut off when they became salaried agricultural workers on Spanish land; and still others fled to the nearby sierra (Henao 1980, 56).

During the nineteenth century the difference between these social classes grew. The class of mestizo merchants and tradespeople in the valley were known as the "people of reason" (*gente de razón*), as the white class (*la clase blanca*), or as "coyotes" by indigenous residents (Henao 1980, 87). Indigenous inhabitants of the valley were referred to as "village folk" (*gente del pueblo*), "the indigenous class" (*la clase indígena*), or simply Indians (*indios*). In some valley towns like Ajalpan these two social classes coexisted within the town, but in the southern municipios of Chilac and San José the populations remained predominately indigenous and mestizos were considered outsiders.

The main sources of water for irrigation and domestic use in the indigenous communities were also usurped by the Spaniards and later hacienda owners. All three of San José Miahuatlán's most important springs, Coyoatl, Cozahuatl, and Atzompa, were used by the population during pre-Columbian times, and were partially appropriated by valley haciendas and the Calipan sugar mill during the seventeenth and nineteenth centuries. The Coyoatl ("water of the coyote") spring was lost to the Calipan mill in the seventeenth century, but a political faction from San José regained control of it during the post-revolutionary period of agrarian reform.

A second important spring for San José, Cozahuatl, is near Chilac's La Taza spring. Springs (and galerías) can be found within the borders of one town while their water flows down the valley slope through canals into the next town, benefiting residents downstream. In the early decades of the twentieth century a water association from San José augmented its use of Cozahuatl by buying water access from the Hacienda San Andrés. San José also relied on the Atzompa spring, whose flow of water was so abundant that it was divided into takes or "waters." At the beginning of the nineteenth century the Hacienda Venta Negra had had control of Atzompa's "Big Water" (Agua Grande or Cuallatl), so named because of its impressive flow of 460 liters per second. Atzompa's waters were the source of many disputes.

The neighborhood system of political organization and water distribution originated during the formation of indigenous villages as units of colonial tribute, and underwent strain as the nineteenth century came to a close. This was a period of hacienda consolidation and expansion. By extending the legal changes brought about by the liberal reforms, the administration of Porfirio Díaz (the Porfiriato) facilitated the division of communal land and the confiscation of idle land or land without title. As a result hacienda owners and speculators were able to buy or seize Indian land and water in the valley, which they used for cultivating wheat, maize, and sugarcane, as well as raising goats for a large annual slaughter.

The two main commercial producers in the valley—the Hacienda Buenavista in Ajalpan and the Hacienda San Francisco Javier Calipam (alternatively spelled Calipan)—had sugar extraction mills, and their operation required water, land, and labor from valley communities. Calipan, for instance, appropriated both land and water from San José (Enge and Whiteford 1989, 79). During the peak of the hacienda system indigenous communities had to share the shrinking water resources, and they did so by increasing the number of cargos and mayordomías in their communities (Henao 1980, 258).

Indian villages also repeatedly struggled to regain their means of production. Their claims were frequently met with the destruction of canals, clandestine diversions of water, the altering of water measurements, and physical threats and violence by hacienda owners and staff (Henao 1980, 66). Disagreements between indigenous inhabitants and *hacendados* were taken to the municipal government in Tehuacán, which was controlled by large landowners (Salazar Exaire 2000, 79). Some indigenous residents of the valley rented water and land from haciendas. Remarkably, residents also pooled their money to buy back water rights and land from hacienda owners, as in San José in 1765 (Henao 1980, 71).

As haciendas looked for new means of access to water resources in the latter half of the nineteenth century and into the twentieth, so too did Indian sharecroppers and peones. It was in this period of hacienda expansion and competition over water that indigenous residents also began to organize modern water associations to share the labor and cost associated with constructing and maintaining chain wells.

Galerías Filtrantes

The valley's galería filtrantes are an impressive and complex tunnel system of chain wells. A galería is an excavated subterranean tunnel that takes advantage of gravity to bring water to the surface at a lower elevation and into canals, sometimes several kilometers away. The length of one galería can reach twelve kilometers, while the entire underground system is estimated to contain 230 kilometers in total (Woodbury and Neely 1972, 143). Each tunnel has a number of *bajadas* or wells that are used to descend into the tunnel in order to complete excavation, or after completion, for cleaning and maintenance. Some male residents of Chilac specialize in the extremely slow, arduous, and dangerous work of excavation. The construction of galerías, like the construction of deep wells, is an expensive and risky business venture, since there is no guarantee that water will be found. For this reason associations are often formed by friends and relatives who share the labor and expense of excavation and construction.

Water associations are collective and locally managed. Previous research on the irrigation system of the valley highlights both the communal nature of the associations and the tensions generated by the accumulation of resources. Associations are both profit-making ventures based on the capitalist principle of accumulation and egalitarian associations structured along the lines of indigenous institutions—with a democratic structure, the annual rotation of leadership, and a patron saint (Enge and Whiteford 1989, 110). Like the ejido and communal land associations in San José, local sociedades consist largely of men. Women make up a small minority of the association membership and are generally widows who have inherited their share from husbands or male relatives, although this was not necessarily so in other valley towns.

Some scholars believe that the galería filtrante was an indigenous invention, but because there is no record of it before the nineteenth century, the regional literature suggests that the Spanish brought the technology to Mexico after learning the qanat system from the Moors (Woodbury and Neely 1972). However, the complex irrigation system in use today is also the result of cumulative indigenous innovations dating back to ancient times (ibid.). The building and management of the galerías is an incredible regional achievement.

The construction of galerías increased rapidly from the mid-1940s to the late 1960s. Although galería associations were a local and regional adaptation to the increasing competition for diminishing water resources, their proliferation, along with that of agribusiness, contributed to the overuse of water in the valley. From 1978 to 2001 the number of working valley galerías declined from 129 galerías reported in 1978 to 75, according to officials with the Ministry of Agriculture (interview, 9 May 2001). In San José residents report using four small galerías (a fifth dried up in 1985) and three deep wells. Groundwater is understood by valley residents to be limited—they have felt the effects of overuse—but nevertheless some associations proceed as though water resources were limitless.

The Atzompa spring was an important source of community water. According to the town elders I interviewed, all adult male residents had usufruct rights to spring water and communal land, but they were required to contribute labor service or a small tax. The mayor's office, or ayuntamiento, collected the tax and oversaw the diversion of Atzompa water to different locations. In exchange for his labor or cargo service each male head of household received an *acción* of six to twelve hours of water per month. The water destined for cultivated land was called "Martesatl" water—a Nahuatlization of *martes*, the Spanish word for Tuesday, when agriculturalists received their water. The Gastoatl (or *Pilahatl*) water from Atzompa provided for the town's domestic water use. Finally, mayordomos, when fulfilling their year-long cargo service, received twelve hours of *Ranchoatl* (or *Atltepeame*) per month, which was enough water to grow the needed maize for the fiestas they hosted in honor of the saint's image temporarily in their care. Each mayordomo and his committee of seven male helpers (and their wives and households) received the twelve hours of water per month during their year of service. Because there were eight mayordomos, a total of sixty-four households received a portion of this Ranchoatl water.

Today the mayor and his municipal council, along with two other committees (the mesa directiva of each barrio and the church committee), are responsible for managing public works and residents' service to the community, town celebrations, and law enforcement.[6] The neighborhood committee is led by a councilperson (*regidor*) who also works at the ayuntamiento, and meets once a month to discuss local matters. Popular assemblies are held to elect the mesa directiva, whose responsibilities last a year. The regidor has the job of informing the mayor about projects that need to be carried out in the neighborhood. The neighborhood committees also choose candidates to fill the mayordomías, who are then asked by the municipal government to fulfill the cargo. Service was formerly obligatory, but the ayuntamiento can no longer require anyone to be a mayordomo, although there remains social pressure to accept the post. Increasingly the value of mayordomías is under debate by residents, and migrants have access to other forms of social prestige, as discussed in chapter 5.

Men and women who perform a service on these committees or fulfill a cargo can vote and are granted a recognized voice to speak and influence opinion at assemblies. Committee service is tied to the accumulation of respect and prestige through the sponsoring of ritual kin (as godparents) and the hosting of fiestas and life-cycle rituals, such as mayordomías and weddings. The more services performed, cargos held, and parties hosted, the more respect residents gain in the community—if those services are carried out well—and the more valued their opinion. Although women cannot be mayordomos, as the wives or kin of mayordomos they help care for the image and prepare the celebrations, and receive social prestige for doing so.

Community service, like the division of household labor, is gendered. Women tend to gain social prestige and respect through service on church committees or as co-sponsors in ritual kinship arrangements, fiestas, or the mayordomía. They may also have important community roles such as healers or *tetlallicihuatl*, the wives of tetlallís who greet and attend to female guests. But residents also suggest that men are the main providers for the family, supplying food or bringing in an income when in fact women may also make such contributions. When I asked Marta, a Sanjosepeño mother in her early forties, whether women could be elected to the mesas directivas of the barrio, she said no. Her explanation was that service on the committee, and similar forms of community service, are closed to women because women do not "work" (*trabajar*). When I pointed out to Marta that she runs a business cooking food for other people, that indeed some women do both paid and unpaid work, she responded: "Well, yes *now* women work, but they didn't before. Things are really different now. Women didn't get past primary school before. Now they get an education, they work. Now there are even women working in the municipal government!" (interview, 28 February 2002). By federal law the mayor's staff must now consist of a minimum of 20 percent women. Although Marta pointed out that women have paid work or employment today, her explanation not only refers to the tradition of the male household head—a tradition found in the local allocation of water and state-recognized land rights (the ejido discussed below)—but relies on the assumption of the male breadwinner. The assumption prevalent until recently that "women do not work" overlooks the fact that in the early twentieth century, on top of the unpaid duties of the household, like shelling maize, preparing meals, and working in the fields, women contributed to household income by selling produce at regional markets. In some cases women were also healers. At mid-twentieth century girls and young women became paid domestic servants in Mexico City, and later took up paid piecework embroidery in their homes in San José. In contrast to unpaid work and tasks performed by women, the unpaid male farming of the milpa in producing subsistence corn (or cash crops) is considered work. Cultivation of the milpa is described as physically arduous, sometimes requiring nighttime irrigation and therefore not suitable for women.

During the first half of the twentieth century the majority of San-josepeños were not socios of water associations but irrigated crops with water from springs and the ravine. Service to the community, such as through the mayordomía or faena, thus provided important avenues for obtaining spring water; it enabled those households without alternative water access to grow corn for household consumption and wheat or tomatoes as cash crops. Additionally, for those Sanjosepeños with landholdings near the ravine (or *barranca*) of the Zapotitlán River, the collected water from the ravine provided an irregular but important source of irrigation water, as it does still. The valley use of this ravine water for irrigation has been called part of the "most extensive and complex runoff management system in Middle America."[7] Ravine associations (*sociedades de la barranca*) from San José and Chilac built diversion dams across the ravine, which push water into a network of canals and into their fields. The water deposits rich topsoil on the fields, helping the productivity of the land. Rain from high elevations in the sierra runs down and floods the ravine with incredible force. For this reason the irrigated parcels (or *pantles*) are surrounded by raised earth borders to withstand the water's impact. A pantle is a parcel of irrigated land which ranges in size from one eighth of a hectare to half a hectare and comes in various shapes.

Sharecropping, or working *a medias*, also provided access to irrigation water for non-socios. Several variations of this basic sharecropping arrangement existed, but most commonly there were three parties: a socio who contributed the necessary water, a second person who contributed land and some labor, and a third person who contributed most of the labor, as well as the fertilizer, seed, and oxen. The three parties divided the harvest equally. Another common sharecropping arrangement was between a water socio and a second person with land but no water.

In the context of growing class disparities between an upper valley peasantry with water and a lower valley peasantry without water, sharecropping arrangements linked households of different means. Moreover, sharecropping involved ritual kinship, whereby the more prosperous peasant would become the patron of his less prosperous counterpart for important life-cycle rituals and celebrations. As previously mentioned, such arrangements bestow respect and social prestige on the patron.

The flow of the Atzompa began to taper off in the 1950s and then dried up completely in 1954. When this happened sharecropping and the renting of *acciones* (or water shares, usually for five or six hours of access) from other valley associations became even more important for those Sanjosepeños left without irrigation water. Non-socios can rent an acción of irrigation water from valley socios, such as at the daily water market in Chilac, but agriculturalists often complain about the expense and rising cost of the water.

When Atzompa dried up, the municipal government found itself without water to compensate residents for their service to the community. As a temporary solution, authorities provided mayordomos with water from the Cozahuatl spring; nevertheless, the practice of the mayordomías was significantly altered. Historically the care of the eight religious images had rotated among the mayordomos in each of the four barrios, but after the water dried up, two images and their associated celebrations were assigned to each barrio. Today the mayordomos are still entitled to spring water, but Cozahuatl produces a flow of only six liters per second; this entitlement to water is more symbolic than real. At the same time, the prestige attached to the mayordomía is on the decline, and residents are becoming more reluctant to undertake the cargo.

The two forms of water access—neighborhood distribution and private associations—ostensibly coexist in San José to this day, but in a much altered form. By mid-twentieth century private associations had become the avenue through which most agriculturalists acquired water above and beyond sharecropping, rainfall, or ravine water. While elements of the barrio system remain today, there is little water to distribute in exchange for cargo service, particularly since the drying up of the once abundant Atzompa spring in 1954.

Don Ignacio's story below illustrates how sharecropping links residents and households of unequal status and means in a patron-client relation of ritual kinship. But as we shall see in later chapters, sharecropping is now on the decline.

Don Ignacio

Don Ignacio, a seventy-two-year-old campesino, remembers that when he was a boy there was no highway to Tehuacán and the trip took five hours by mule. Like most boys he began to work the milpa with his father when he was twelve or so. His family also grew wheat to sell to distributors from the neighboring towns of San Sebastián, Chilac, and Altepexi. In his early twenties Don Ignacio worked for a couple of seasons as a cane cutter in Veracruz, as did many male Sanjosepeños. In 1955 he became an agricultural worker in the United States under the Bracero program and saved enough money to buy several hectares of land in San José. Don Ignacio was one of the few dozen men from the municipio who had worked in the United States as contract workers.

His father was a socio of the Atzompa water spring, but he did not leave Ignacio a water share because the spring dried up in the 1950s. Instead Don Ignacio practiced a medias for two or three cycles a year, putting in the necessary work while a friend who was a water socio contributed the irrigation water. This friend later became a godparent to Don Ignacio's children. Don Ignacio also saved and bought a share of the San Agustín water (a nighttime share of Atzompa) with some of the money he made working in the United States, but sold it before the violence of the 1980s started. A few years ago Don Ignacio stopped growing crops for sale because, he says, it became too expensive. He now grows a few small pantles of corn for his household and his adult daughter. For Don Ignacio, like many others, sharecropping was a key relationship in the distribution of water and labor between households. Wages from agricultural labor and cash crops were also needed to purchase water or, as the population grew, land.

Landholdings

In the early twentieth century residents owned private, individual landholdings or used community land administered through the mayor's office. In 1946 San José also received an ejido of two

TABLE 2: Labor and Land in San José Miahuatlán, 1970

Type of Landholding	Total Number of Units	Number of Producers and Family	Number of Seasonal Workers	Number of Permanent Workers	During Harvest of Winter 1968–69	During Harvest of Spring–Summer 1969
Private plots, more than 5 hectares	11	23	3	1	12	24
Private plots, 5 hectares or less	259	356	82	—	88	253
All private plots of land	270	379	85	1	100	277
Ejidos and "agrarian communities"	4	315	157	1	1,231	3
Total	274	694	242	2	1,331	280

Source: Adapted from DGE, *Censo Agrícola-Ganadero y Ejidal*, 1975 Puebla, 14, 356.

thousand hillside hectares of land expropriated from the former hacienda of Axusco. The *ejidal* land tenure system combines communal title vested in the state with either communal or individual usufruct rights to work the land. Some 515 of San José's 1,180 inhabitants obtained rights to the ejido as either heads of household or single men over the age of sixteen.[8] Over thirty women were included as *ejidatarios* in this original list, which suggests that either they were considered household heads or perhaps, as with the Coyoatl spring, men wrote the names of female relatives in documents to gain more household access to the resource. As hillside land the ejido was designated for "communal use," which is to say that it was never divided into individual parcels. Farmers employ the land mainly for goat grazing, collecting firewood (*leña*), and some irregular milpa and string bean (*ejote*) cultivation.

In addition to ejido land San José today possesses 1,200 hectares of state-recognized communal land. In 1973 the federal government officially recognized the *bienes comunales* and approximately four hundred residents became communal landholders (*comuneros*). Additionally, four hundred hectares of communal land reserved for common use (*uso común*) were not parceled up individually. Before official recognition in 1973 communal land had been administered by the municipal government, with the households cultivating the same fields year after year. Any household could clear a parcel of the land for use and pay a small harvest tax to the mayor's office. Those residents with access to communal land today belong to the families that had cleared the land in the 1970s at the time of official recognition (see table 2 for a breakdown of landholdings in 1970).

Under the land titling program (Procede), which began in 1992 alongside other neoliberal reforms, communal and ejido land can now be sold or leased if approved by the local agrarian committee. Currently the communal land committee of San José has an internal rule that the land cannot be sold, only transferred through inheritance, although this could change in the future.

WATER DISPUTES

Sanjosepeños struggled to win back water access that had been previously usurped by haciendas and sugar mills in several ways. Residents petitioned the post-revolutionary state for water (1920s–1940s). This was a state-sanctioned and institutionalized means of seeking redistribution, redress, or mediation for disputed resources; and success often required forging alliances with political bosses or powerful groups in the valley and beyond. Sanjosepeños also formed water associations to combine funds and purchase acciones of the flow from a natural spring or to construct galerías filtrantes. While this was a legitimate means of acquiring water rights in the valley and remains so today, it has never been officially sanctioned at the regional, state, or national level; groundwater is after all the property of the state.

And finally, a third strategy employed by Sanjosepeños was the use of force. Local political factions employed violence against valley agriculturalists who challenged their claims to a particular source of water, and in some instances against a different political faction from San José vying for political influence and control of the precious resource. In regional newspaper accounts from this period rural violence is depicted as a savage Indian trait, a depiction that clearly draws on colonial distinctions between the people of reason (gente de razón) and Indians.[9] But the threat and use of violence was not unique to San José: the formation of the valley peasantry entailed the use of force by the state and hacienda owners in staking claims to the resources of the valley.

During the revolution and the following period of agrarian reform, political factions fought for control of the Tehuacán Valley and its resources. While a highland or *serrano* movement was active in the north of Puebla state fighting to regain local autonomy from outside political centralization, an agrarian (*agrarista*) movement focused on land redistribution emerged in the central highlands and south (Pansters 1990, 39–40). Revolutionary struggle was led by medium-sized landowners from the valley and the adjacent sierra (Henao 1980, 127). General Francisco Barbosa of Ajalpan commanded the winning faction and became the most important revolutionary leader in the region. Although Barbosa supported the official post-revolutionary government position, he was more interested in political and economic domination than in any real reform. He was an "entrepreneurial cacique," a powerful political boss who acted as a broker between campesinos and large haciendas during agrarian reform (LaFrance 2003, 42). General Barbosa was murdered in 1925, but his family continued to dominate the region for many years and in the process amassed a good deal of water resources.

In San José the regional factionalism of this period exacerbated existing fault lines in the town, where two main political groups fought over the Atzompa spring and for control of the municipal government. The smaller faction of thirty or so campesinos was led by Alfredo Romero and affiliated with the Tehuacán baker-turned-leader of the Eastern Socialist Party, Ernesto Valerio. Romero was one of few mestizos in San José. His family was the wealthy owner of the only big supply store in town. The other faction, reported to have initially had the support of the majority of the town's residents, was led by Miguel Correo and aligned with General Barbosa (interview with Don Carlos, 22 March 2002). Yet although Correo was linked to this important regional leader, San José was in a weak political position compared to valley neighbors like Ajalpan and Chilac.

Soon after the Atzompa and Coyoatl springs were declared national state property, the governor of Puebla provisionally granted part of both springs to San José Miahuatlán in 1932, based on written requests made by residents.[10] Four short years later this decision was rescinded, even though San José claimed ancestral use before the era of haciendas.

According to the official record, the agrarian authorities decided to give the water to the ejido of the neighboring municipio of Chilac rather than to San José, which had no ejido at that time. Based on the official explanation for the decision to rescind, the agrarian authorities gave preference to water requests submitted by those who had received land (ejidos and bienes comunales) under agrarian reform.[11] In the southern valley both water and land grants were an avenue for generating political loyalty to the ruling party. Elders also suggested to me that successful petitioners during agrarian reform had some degree of political connection to the ruling party. In San José the land irrigated by Atzompa and Coyoatl was mostly privately owned, and it was not until the 1970s that a part of the land irrigated by Atzompa was officially recognized as bienes comunales, or communal land.

The Coyoatl Spring

According to the tetlallí Don Carlos, the Coyoatl spring, which had been originally used by Sanjosepeños, was exchanged for a loan from the Calipam hacienda (and later sugar mill) in the seventeenth century. Since San José was short on funds to finish construction of the local church, residents struck a deal with the hacienda owner. Don Carlos argues that because the deal was oral and never reduced to written documents, Calipam was able to retain access to Coyoatl's water into the twentieth century (interview, 22 March 2002). Not only was it difficult for residents to prove their original access to the spring, but when other valley communities were petitioning the state for land and water during agrarian reform, Sanjosepeños were unsure how to proceed. I asked Don Carlos if San José had bought back the water from the Calipam. He replied:

> No, not bought, reclaimed because the water is from here and since it was lost, that is to say, the government, well, the revolutionaries won and they were redistributing it during agrarian reform. But our ancestors didn't know how [to reclaim the land and water]. And around that time, the administrator of the Calipam sugar mill, Don Severino Carrera Peña, spoke to Miguel Correo [the mayor of San José at the time] and said, "When you get [ejido] land I'll give you the water." Don

Severino said, "I'm not the one who controls Coyoatl now. Now the ejidatarios [from the nearby towns] control the water. You can make a claim. What are you waiting for?" The residents from Calipam, Coxcatlán, and San Sebastián took advantage of the situation, they solicited and received the documents as ejidatarios and got water that had been used by the sugar mill. (interview, 22 March 2002)

Don Carlos suggests that Sanjosepeños often lost to their neighbors in the scramble for claiming land and water because of their lack of education and political connections. The cultivation of alliances and the role of education, especially the ability to speak and write Spanish for the production of documents, was a salient feature in such interviews with elders. In effect residents articulated to me that for their claims to be acknowledged by the state, they had to adopt its language and procedures (which included cultivating the right connections), something that was often difficult to accomplish. Also, in several important claims to water the educated used their skills to benefit their own political faction.

In 1942 the mayor Miguel Correo and his political faction helped to organize more than a hundred Sanjosepeño men and women armed with sticks and machetes to take over the canal that ran from Coyoatl to the ejidatarios in Coxcatlán. Don Carlos explains how residents took up arms to take back their water and so that no one would "bother them": "It cost us [Sanjosepeños] a lot of work with guns and everything to get Coyoatl back. It's a shame. . . . but [our parents] took up arms so that no one would bother them" (interview, 22 March 2002). When petitions and regional political leaders failed them in reclaiming their ancestral water, Sanjosepeños took over the canal by force. With Coyoatl back in the hands of Sanjosepeños, Correo's faction attempted to claim the water as its own, but tensions mounted. The members of Correo's faction did not own any of the land connected to the Coyoatl spring by irrigation canals, and so were unable to justify controlling the water to their fellow residents. The ayuntamiento was forcefully taken over by one faction, then another, and then still another, with which the water finally stayed and developed into the "property" of a water association.

Although Miguel Correo was displaced from political office, other members of his family later emerged as politically important in the municipio. Correo's distant nephew Felipe would become a cacique in San José, owning land, water, and a soft drink and grocery distribution business serving the valley. He also was a central figure in the Atzompa conflict discussed below.

In the late 1940s a small delegation of educated, bilingual men from the faction that had taken over Coyoatl and the municipal government went to the Ministry of Water Resources in Mexico City to petition for water.[12] In 1954 the water society of Coyoatl from San José—made up of 121 water users (usuarios)—was legally recognized by the Agrarian Commission, a development which buttressed this group's power within San José.[13] With help from two important national leaders of the National Confederation of Campesinos (CNC) who lived in the region, the agreement was ratified in 1966. The different towns of Calipan, Coxcatlán, and San José county all respected the town's claim to the water.

While in other areas the regional committees of the Ministry of Agriculture and Water Resources (later split into two branches, the Ministry of Agriculture and the National Water Commission) had representatives from campesino organizations, this was not so in the valley. The National Confederation of Campesinos (CNC) was weak in the region, dominated by political bosses who profited from the private ownership and accumulation of water. The CNC did not support the nationalization of water in the area (Enge and Whiteford 1989, 177).

This Coyoatl group also petitioned and received an ejido of two thousand hillside hectares of land designated for communal use from the former hacienda of Axusco in 1946.[14] Today the two Coyoatl organizations, the water association and the ejido, are interrelated: each has its own committee, but all the important decisions —other than those specifically related to the ejido—are made by the Coyoatl water association committee.

In the twenty or so years between the return of the Coyoatl waters to San José in 1942 and the ratification of this return in the 1960s, at least ten people were killed in town disputes connected to the spring. Of the different factions that clashed in the struggle over Coyoatl and the ayuntamiento, the one that came to control Coyoatl (now the Sociedad de Coyoatl) and the municipal government in the 1940s was the most politically powerful group in both the town and the municipio. Through the work of several bilingual and educated members it made connections with leaders from the National Confederation of Campesinos. Until the Atzompa murders of the 1980s most of the mayors of San José came from within the ranks of the

Coyoatl water association. The most important rival to the Coyoatl water association during this period was an association that controlled the Atzompa spring, made up of some members from Miguel Correo's faction.

Atzompa

The government decision to withdraw access to Coyoatl was simultaneously a retraction of access to two waters (San Agustín and San Lucas) of the Atzompa spring. These waters had been requested and temporarily given to San José, but in June 1936 the government retracted its earlier decision and granted the water to the newly formed ejido in Chilac.[15] Later that year a group of armed Sanjosepeños took control of the irrigation dividers that sent the water to the different villages. Troops were deployed after residents from Chilac asked the Tehuacán authorities to intervene. They captured six Sanjosepeños while the others escaped. The influential agrarian leader and head of the Chilac cane cutters' union, Melitón Ramírez Valdez, who had successfully negotiated the granting of a share (*dotación*) of the Atzompa waters for the ejidatarios of his town, was shot dead by a man from San José (Ortega García 2001, xxi). Members of the union blamed his murder on San José, but they also suspected a network of caciques who challenged the agrarian movement's attempts to form ejidos (Enge and Whiteford 1989, 104).

In 1937 a water association from San José bought access to the Atzompa spring. In other words, the association did not receive recognition of a water claim, as had been done in Coyoatl, but rather purchased access.[16] This group of thirty Sanjosepeños bought the rights for fifty years of access to Atzompa's Agua Grande and San Agustín waters from the hacienda Venta Negra, the same hacienda whose land formed the new ejido in Chilac.[17] Later, in the mid-1950s, the San José association would enlarge its share of Atzompa by buying access to more of the spring's water flow.

With the increasing construction of galerías filtrantes and deep wells higher up the valley the water table began to decline, and the force of Atzompa's water flow began to taper off. The town was left without a water source for the mayordomías, although another spring, Cozahuatl, provided provisional compensation. In an effort to remedy the diminishing supply of water, the association began to excavate for a galería in 1948. When the spring water dried up six years later, in 1954, the ayuntamiento decided to join forces with the association by contributing labor and finances to the excavation of the galería, in the hopes of locating and tapping into more Atzompa water.[18] By this time the association's leader and mayor was Felipe Correo, the distant nephew of the former mayor Miguel Correo. After a few years of contributions, however, the ayuntamiento (with a new mayor) withdrew its share of resident labor and funds from what seemed to be too costly and risky a venture.

Adult Sanjosepeños remember where they were and what they were doing when they found out that Atzompa's waters had returned in 1975. With the welcome news, the municipal government began to dedicate approximately 20 percent of San Jose's total access to the Atzompa for the town's domestic use. The rest of Atzompa's flow went to the association of thirty-two socios to irrigate their fields.

In practice, socios purchase shares (acciones of six hours per month) and pay upkeep fees to the association. Those who cannot pay their fees must rent or sell their shares, enabling others to become association members or accumulate shares. By the 1970s Felipe Correo owned shares in all the important spring and galería associations in San José, including Cozahuatl, Coyoatl, and Atzompa. Don Felipe became the majority shareholder of Atzompa, owning between seventy-six and eighty hours of water. He was a *mediador*, the owner of land or water in a patron-client relationship of sharecropping. In the nearby municipio of Ajalpan those socios who owned 50 percent or more of water shares dominated association decisions (Henao 1980, 150). Don Felipe was a similarly dominant figure in the water associations and politics of San José.

By 1980 the discord in San José over who was entitled to the majority of Atzompa's water had escalated. Don Felipe and the water association claimed the rights to Atzompa, based on the money and labor they had invested into the galería, while the municipal government argued that the water, as part of the public domain or national property, was really to be used by the entire community. When I asked about Atzompa one campesino told me, "The water association started to get water with eight or ten meters of galería and that's when the problem started. It split the town in half. Some were saying that the water belongs to the government, that it was the property of the nation" (interview, 22 February 2002). Residents held distinct ideas about whether the Atzompa spring water was the property of the association or rather public property. Did the water belong to the members of the association, or was it to be made available to the entire community of residents? The declining rates of water flow in the southern valley springs contributed to a change in the local distribution of water. The tension between the older (barrio) system of water management and private associations, and distinct ideas about what sort of property water was, erupted in the case of Atzompa. A central fact was that a local group had aligned itself with a new and influential party in the valley, the PSUM (the Unified Socialist Party),[19] which had put the unequal distribution of water in the valley on the regional political agenda. When local PSUM members took over the municipal government in the early 1980s the water dispute intensified. The new members of the ayuntamiento also invited state-level leaders of the PSUM to accompany them to the site of the Atzompa spring, in an attempt to take the water away by force from the association. As one resident explained: "In those days, everyone was with [the national ruling party of] the PRI in San José. The PSUM entered with issue of Atzompa and the [conservative party] PAN arrived at the end of the 1970s. PSUM represented 'el pueblo' in the taking of Atzompa, but people saw the PSUM behaving badly after this, and many people changed back to the PRI. Water socios were killed by the [local] PSUM, and then they began to fight and kill amongst themselves!" (interview, 6 March 2002). PSUM representatives had used the issue of Atzompa to gain entrance into San José's political affairs, but once involved the local members recreated older feuds and alliances. Although official PSUM policy advocated against the private ownership of water—an unpopular stance in the valley—in San José party members sought to enhance their own water holdings and those of their allies. In this way the conflict in San José had more to do with the local working out of state-led agrarian reform decisions and less to do with official political party platforms.

During the 1970s and 1980s opposition parties had a real presence in the district of Tehuacán, particularly the conservative opposition party, PAN. During the 1980s the PAN remained in power in Tehuacán and some other important urban areas, while the PRI strengthened its power in the countryside, particularly with the help of the peasant organization the Antorcha Campesina. This organization joined the CNC (part of the PRI) and channeled state funds to rural, indigenous towns for roads, schools, and other projects. So while the PRI remained popular in the countryside, the PAN gained popularity among the Pueblan entrepreneurial class because it was critical of the nationalization of banks in 1982 by the PRI national government. In the fraudulent state elections of 1983 it appears that the PAN accepted defeat "in return for victory in the two medium sized cities of Teziutlan and Tehuacan" (Pansters 1990, 160). Although the leftist party PSUM was not a major player in the state elections of 1983, it was active in the Tehuacán Valley.

In San José the PRI was elected to the mayor's office and council in 1981. But in the following year the faction aligned with the PSUM seized power, and the PSUM remained in the ayuntamiento for years. Previously whichever faction held office had always been affiliated with the PRI, but at the local level the Atzompa dispute changed this. Despite rhetoric about water as state property, the Sanjosepeños associated with PSUM who had taken over political office were interested in the spring, not for the benefit of "el pueblo" per se but for those residents aligned with their faction. The disagreement turned violent on 3 November 1981, when the cacique Don Felipe and four others were assassinated by members of the PSUM ayuntamiento and its supporters.[20] A state of emergency was declared in San José after the attack, and for at least two intense years San José was considered a "lawless town."[21] Tensions within the ruling faction escalated and led to more bloodshed.

For close to two decades residents who were not affiliated with PSUM lived in fear of the municipio authorities and their police. In 1999 the PRI managed to oust PSUM at the local level for one term. Then in late 2001 the PRD (whose membership came from the now defunct PSUM) won the local elections, some claim fraudulently. For the most part the tension has dissipated, but old animosities and divisions still affect everyday life in San José; strong PRI supporters frequent stores owned by PRI residents, for instance, while PRD members shop at PRD stalls and stores.

In San José conflict over water was not the result of national party policies but rather of the ways local representatives of the parties took sides in local disputes and the extent of their economic and political resources. In a town the size of San José, local feuds between political parties often become feuds between families. In his research on rural violence in the adjacent state of Oaxaca, James Greenberg (1989) argues that what begins as a dispute between individuals often grows, as each new murder brings more families into the conflict and produces new enemies. In San José, if someone is murdered a male kinsman is expected to avenge the death. When I asked about how Don Felipe's soft drink business fared after he was murdered in the Atzompa violence of 1981, I was told that most of his children were in the United States working. They had left in the mid-1980s, "because they gained the reputation as cowards for not avenging their father's death" (interview, 10 March 2002).

Greenberg also points out that "the dense crisscrossing of kinship and friendship lines within the village also serves to check" the expansion of blood feuds (1989, 222). These relationships do sometimes restrain further escalation of the feud. Interested parties may uphold civility in the name of these relationships while hoping for revenge. But these situations are still characterized by a general atmosphere of distrust. This was the atmosphere in San José in the 1980s and 1990s.

In the 1980s the state "unofficially" intervened to help resolve the Atzompa conflict. The regional office of the Ministry of Agriculture bought Atzompa shares from the San José association, but by some residents' accounts it was too little, too late. I was told by association members that the government did not admit its involvement but "paid socios to calm things down." In effect the buying of shares was an attempt to convert the water from private property of the association to public property of the national state.

Even when residents in San José blame the water association and not the faction in office for the Atzompa conflict, most defend the principle that water generated through investment is the rightful property of the association collective and its members. Although state participation could produce a more equitable distribution of water, many valley residents reject that arrangement in favor of ownership based on investment (interviews, 3 March 2002; Enge and Whiteford 1989, 185).

In recent years neoliberal policies have decentralized irrigation management, and while this has ironically entailed an increase in the state's role in irrigation management of the Tehuacán Valley, residents continue to rely on the regional system of water associations and the renting and selling of water shares.[22]

THE LANGUAGE OF CONTENTION AND
BEING A "DAMNED" INDIAN

We are being punished by God . . . We are being punished
for all the violence. More blood has flowed [here] than water.
—Indigenous campesino from San José, 10 November 2001

Those damned Indians! Violence is all they know!
—Indigenous campesino from San José, referring to the
history of San José, 16 February 2002

The Atzompa murders are a defining moment for many adult Sanjosepeños in that the events of their lives are explained in relation to
the violence and the drying up of the spring. Social memory of
the violence, along with that of previous water conflicts, shapes
residents' images of themselves as members of a community and
an Indian town. Many socios and residents on either side of the
Atzompa conflict explain that the excavations of wells and galerías
further up the valley adversely affected the water levels supplying
San José. Yet despite this explanation, residents also commonly at-
tribute the drought to divine punishment. This sense of being punished is interwoven with the sense of being an Indian.

Throughout the region Sanjosepeños have the reputation for being fierce and warrior-like (muy guerreros). They are sometimes referred to by neighbors in the city and valley as "people without
feeling, without emotion"—in the words of one middle-class mestizo resident of Tehuacán—or as indios and savages. Sanjosepeños
negotiate this reputation in different and sometimes contradictory
ways. Some interviewees suggested to me that when their fore-
fathers' petitions for water proved unsuccessful, the mobilization
of violence against outsiders was their "only alternative." They regarded violent conflict over water as resulting from regional political battles to control resources, or from outside attempts to usurp
property and subordinate residents; in this view, their reputation as
fierce warriors emerged as a means of defending their claims to
water resources. Other residents whom I spoke to held multiple,
contradictory views about the town and its history of violence. When
I asked about the Atzompa violence, one indigenous Sanjosepeño in
his mid-fifties referred to his fellow residents by exclaiming, "Those
damned Indians! Violence is all they know!" With these two short

sentences he communicated how tightly bound the experience of violence is to being an Indian, as well as the idea that some individuals or groups are more Indian than others.

Notions about being a Sanjosepeño today emerged through a history of negotiations with the state, including the colonial project and the "state activities, forms, routines and rituals" of the postrevolutionary period, such as water petitions and the mobilization of violence (Corrigan and Sayer 1985). William Roseberry has pointed out that the "forms and languages of protest or resistance must adopt the forms and languages of domination in order to be registered or heard" (1994, 364). "To the extent that a dominant order establishes . . . legitimate forms of procedure, to the extent that it establishes not consent but prescribed forms for expressing both acceptance and discontent, it has established a common discursive framework" (Roseberry 1996, 81). As we have seen, groups in San José formulated their claims to water and land through the procedures of agrarian reform. They also negotiated and deployed earlier, racist stereotypes about the Indian proclivity for violence. As previously mentioned, regional newspaper accounts from the early twentieth century portrayed valley Indians, and Sanjosepeños in particular, as brutal savages.

This is not to suggest that state rule is the hegemonic accomplishment of consent or a shared ideology, nor that Sanjosepeños simply internalized racist stereotypes about Indians. Rather, notions about being a Sanjosepeño emerged as part of a contested process of constructing "a common material and meaningful framework for living through, talking about, and acting upon social orders characterized by domination" (Roseberry 1994, 361). Sanjosepeños wrote petitions during agrarian reform, established water associations, and used force, in the process mobilizing discourses about violent Indians. This common framework of "community" has meant negotiating state forms and discourses about campesinos and Indians.

Moreover, the conflict over water was not simply external to the boundaries of San José, with the town acting as a united front against outside interests; the struggle for water reconstituted relations within the community and even the meaning of being Sanjosepeño. Community identity in San José—what it means to be a Sanjosepeño—is not simply the expression of shared interests but the product of debate and negotiation over the meaning of community itself (Smith 1991, 195).

While some Sanjosepeños refer to their Indianness in relation to violence, others counter their reputation as violent Indians by emphasizing their humanity. This was made evident to me during a church ceremony for the Coyoatl water association. Before the ceremony I was introduced as an anthropologist studying San José and its customs to a visiting priest from Oaxaca who was born and raised in town. He was about to deliver the annual service for the patron saint who serves the Coyoatl association. During the sermon the priest spoke about not judging too harshly the association's history of violent conflict: "While it is good to have diverse opinions, it is important to overcome arguments and become a model for a divided pueblo. This water association can overcome divisions and diverse opinions. The errors of the past do not need to be repeated. If someone were to write a history of our water association, if there were a consciousness of this history, they would encounter errors committed by the group. Why? Why has the association committed errors? *Because its members are human. We have committed errors because we are human*" (18 March 2002). Without explicitly mentioning the regional discourse about the town which dehumanizes Sanjosepeños, or wider discourses about Indians, the priest is clearly referring to them. By saying that it is human to err, and that like anyone the socios have made mistakes, the priest is in effect asking the audience (myself included) to reject discourses which dehumanize or denigrate the Indian town.

Other residents discussed violence—particularly violence directed at outsiders—in terms of injustice and lack of education. The tetlallí Don Carlos's story about the Coyoatl spring, for example, suggests that in the absence of justice, residents used their reputation as *muy guerreros* as a strategy in dealing with outsiders, be they neighboring towns or regional state representatives, so that they would not "bother" residents. On some occasions residents themselves deploy the narrative about Indian savagery as a way of saying, "Don't mess with us!" or to challenge attempts by nonresidents to usurp their resources. For this reason local narratives about being punished by god or being fierce and violent Indians are not simply an internalization of racist ideas about Indians. These varied narratives should be seen as part of the "material and meaningful framework" of community; they have been shaped through engagement with wider forces, particularly in making claims for resources on the state.

In the neighboring state of Oaxaca, Greenberg found residents articulating a similar argument: that they turned to violence to defend their resources against outsiders or because there was no justice (1989, 147–51). While residents of the Juquila district in Oaxaca explained outwardly directed violence as a reaction to exploitation and injustice, they generally characterized inwardly directed violence as the product of individual personality traits and conditions, especially jealousy, machismo, and alcoholism, or of a culture of "individualism," of self-advancement and greed. Greenberg argues that in an area with a high murder rate the focus on the individual as an explanation of internal violence "directs attention away from the material underpinnings of contention by transforming conflict over the right to use and distribute resources into sexually charged matters of manhood and honor," particularly in discussions about machismo and jealousy (1989, 209). While the parallel between the Juquila district and the Tehuacán Valley should not be taken too far, there are important similarities. I heard violence in San José described by residents not only as the last resort in resisting outside encroachment on resources but as the result of ignorance; individualistic, self-serving, uncooperative behavior; and Indianness, particularly for men.

In contrast, some sympathetic outsider descriptions of San José point to a failure on the part of the regional and national government to intervene to raise the standard of living or stop the massacre of the 1980s. One indigenous agriculturalist and small businessman in Chilac answered my question about the water disputes by saying, "Those damned Sanjosepeños! They don't have water, that's why they have problems! Here we built *galerías y pozos*. We've developed, we've progressed somewhat. We have more water associations. There in San José they are more individualistic, they didn't work together. Also, in Chilac we've had more help. Our people are more educated. They knew how to ask for permission to build pozos and galerías. In San José it was a government failure. They didn't help and they didn't step in to distribute water more fairly when people were saying, 'This water is mine!' There's been a lot of violence there, a lot of desperation" (interview, 25 May 2001). This informant highlighted as causes of the Aztompa violence the lack of state help and education, as well as individual greed.

Although individualism as an explanation of local violence may obscure the material underpinnings of tension and its relationship to wider forces, as Greenberg suggests, individualism in San José is also reinforced by the experience of living in a very tense, divided community. During a conversation about local conflict over water, one migrant explained to me that when he traveled north to the border he passed through similarly arid parts of Mexico, and yet unlike in San José all the people seemed to have electric pumps for their irrigation wells. He told me: "They have more support from the government in the north." But as he was speaking, an older campesino interrupted to add, "The government doesn't help us, but the people here haven't joined together to ask for help" (interview, 6 May 2001). This older campesino saw the conflict as a failure of both the national government and townspeople to work together despite lingering tensions. In a divided town distrust and fear were obstacles in many attempts to work together across political factions and feuds.

THE COMMUNITY AT MID-CENTURY FORWARD

The picture that emerges of San José at mid-century is of a turbulent agricultural community with differential access to water and land resources, frequently embroiled in water disputes. Male service to the community was compensated with irrigation water, but many residents also practiced sharecropping in an effort to secure water. The system of sharecropping, practiced to a lesser degree today, exemplifies how the household in San José is embedded in social networks and obligations, as well as differential claims to resources and labor. These labor arrangements are part of a broader system of patron-client relationships that often developed into ritual co-parenting (*compadrazgo*).

Sanjosepeños cultivated private landholdings, as well as ejido and unofficial communal land; they grew subsistence corn and cash crops like wheat and tomatoes, and worked as sugarcane cutters, market vendors, domestics, and bakers. Male residents found employment as sugarcane cutters in the valley and in the neighboring states of Veracruz and Oaxaca. In the 1950s and 1960s, during the Bracero Program, up to three dozen went to the United States as contracted short-term agricultural workers. At the same time many unmarried young women and girls began to work in middle-class homes of Mexico City as domestic servants, returning to San José for marriage.

The most important commercial crops in San José were wheat and tomatoes. In the southern parts of the municipio residents also grew sugarcane to sell to the nearby Calipan mill. San José was a wheat-producing county until the 1970s. Today the only sign of the crop's former importance is to be found in old houses, whose walls were made from adobe mud mixed and strengthened with the waste from the wheat stalks. Sanjosepeño cultivators sold their wheat to regional mills, and local bakers purchased their flour from these mills to make their specialty breads. San José is known for its *pan de burro*, so named because bakers would load their burros with bread and journey to other towns and into the sierra. This bread was exchanged for goods like coffee, which they would bring back with them. It took five hours on a burro to ride to Tehuacán, or two days walking to Huautla, Oaxaca, where pan de burro was also sold. In the 1970s such trips were significantly shortened when a highway was built connecting San José to other valley towns.

Although wheat dominated commercial agriculture, residents grew rain-fed corn for consumption and petty sales. In both San José and Chilac some varieties of maize are cultivated today that are the same as the traditional ones, at least in name and according to their basic characteristics; these include *nahuitzi* (rain-fed corn with wide kernels and a four-month cycle) and *macuiltzi* (a sweet, wide-kernel corn of four and a half months).

The area underwent other important changes during the middle of the century, including the introduction of electricity in the 1960s. The increase in population was also a strain on the available land and water resources. At mid-century the population of the head town of San José was reported to be close to 1,180.[23] Twenty years later, in the late 1960s, according to a former mayor whom I spoke to, the population had more than doubled, reaching almost 4,000. By the year 2000 the population was close to 7,000, and agriculture was one of many livelihood strategies, as residents further diversified their means of generating an income.

Ironically, for many producers government agricultural help arrived in San José rather late, after the local crisis was exacerbated by the national economic difficulties of 1982. Sanjosepeños were the recipients of rural credit and agricultural development programs for the first time in the years just before the restructuring of development institutions in the 1980s. According to an extension officer at the National Indigenist Institute (INI) who had worked in San José in the 1970s, their office introduced the first such programs in the county. INI was the point of entry for other government agencies like the National Rural Credit Bank (Banrural) in San José (interview, 26 February 2002). The INI office was working on programs to modernize farming practices such as introducing tractors and fertilizers for corn production, and on setting up a schoolhouse, when conflict in the head town became violent. Some residents relied on Banrural for agricultural credit during the 1970s, and a CONASUPO store was opened in the 1980s. These small INI, Banrural, and CONASUPO projects amounted to the only direct state role in Sanjosepeño agriculture.

San José has been profoundly shaped by struggle over water resources. Access to water was organized around service to the neighborhood and village community, but it was also shaped by, and helped shape, relations of prestige, class, and gender. During the twentieth century this neighborhood system of water access was increasingly undermined as water levels declined and water associations gained popularity in the valley. One of the main sources of water, the Atzompa spring, disappeared at mid-century, affecting agricultural relations and the religious and civil cargo system. However, rather than see this as a "tragedy of the commons"—the result of mismanagement or overexploitation by self-maximizing individuals of communal or open-access resources—I argue that Sanjosepeños creatively, and at times violently, responded to the usurpation of their resources and diminishing water levels in several key ways, including by banding together to form private and collective water associations.

The struggle over water between San José and other valley communities was affected by San José's relatively weak position vis-à-vis regional political leaders and the state agrarian authorities. When the national state did not recognize or honor San José's water claims advanced by petitions during the height of agrarian reform, residents resorted to outwardly directed violence to take their water by force. But factions within the town also clashed over controlling the mayor's office and the Atzompa and Coyoatl springs. The Coyoatl association seized control of the water and the municipal government, and then returned to the business of petitioning for official recognition. They also cultivated an alliance with CNC representatives and won state recognition for their claim on the spring.

As for Aztompa, control of the spring was also accomplished through force and the private purchase of water shares. But problems resurfaced with the construction of the Atzompa galería. When the ayuntamiento and the association contributed labor and funds to construct the galería, followed by the ayuntamiento's withdrawal shortly after, the question of ownership and community access was thrown open to contestation and violent conflict. By the 1980s local factions were expressing their disagreement over water ownership through membership in competing political parties, the ruling PRI party and the leftist party PSUM, which had only recently become active in the valley. Conflict over water and a second Atzompa drought reinforced Sanjosepeños' reputation as violent Indians, and became factors in the outmigration of residents during the 1980s. Today local explanations of the Atzompa-related murders are often contradictory in their negotiation of the wider discourses about a violent Indian character. Violent conflict, factionalism, and the language of divine punishment became part of the "meaningful framework" of community in San José. At times tensions still run high.

Sanjosepeños were grappling with this violent conflict, a water shortage, and related agricultural problems when crisis-related inflation hit in the early 1980s. As a result of these combined pressures, some residents abandoned wheat and took up commercial elote corn production, while others left in search of employment in cities and in the United States, or found other off-farm employment. In the following chapters we will see how residents adapted to economic and water crises, intensifying the need for off-farm income and contributing to changes in the practices of maize agriculture and attitudes about it.

Chapter 4

REMAKING
THE COUNTRYSIDE

Just as capitalists have responded to the new forms of economic
internationalism by establishing transnational corporations,
so workers have responded by creating transnational circuits.
—Rouse 1991, 14

In times of crisis, when social services collapse or cannot effectively
carry out their functions, corn's importance becomes self-evident.
Recourse to corn is the last line of defense for security, for hope, for
the retreat of lesser units of society in order to defend their
very existence.—Warman 2003 [1988], 20

Some of us grow corn because there is no other work. Not everyone
can get a job or make it across the border.—Maize farmer,
21 June 2006

On one of my walks past the cinderblock houses of San José to the
fields, where the paved roads turn to dirt, a campesino in his late
fifties showed me the land where he used to grow wheat and now
cultivates elote, a variety of corn harvested for corn on the cob. Like
many other farmers, Don Modesto has planted several cornfields
this year. Two will be ready soon when elote prices are high, and a
third, to be planted during the rainy season in June, will be harvested
in October. As we approached his land we saw footprints on the
borders of his milpa. Potential buyers from a neighboring town had
come to see when Don Modesto's elote would be ready. But a couple

of weeks later, when no buyer approached him about selling elote, he decided to leave his corn to dry in the fields so it could be used as grain by his family. A small portion of the milpa was also sold to neighbors and relatives.

Don Modesto told me then that not all households have milpas, because there is not enough land for every child to inherit a parcel from their families (interview, 10 February 2002). Further, he explained that agricultural day labor is becoming more expensive: "We used to work like mules when we worked the fields. Now the young ones want to be paid by the hour." Don Modesto believes that members of the younger generation are accustomed to being paid by the hour because of their work experience in valley agribusiness, maquiladoras (or maquilas for short), and the United States. But he adds, "They just don't want to work the fields anymore." Indeed, as another farmer quoted above suggests, growing corn is what residents do when they are unemployed or cannot "make it across the border" (interview, 22 June 2006).

Although corn agriculture is a preferred livelihood for some, this chapter examines how it is increasingly seen as secondary or supplementary to off-farm employment, particularly in the United States. I argue that neoliberal globalization has both generated a larger, more flexible workforce among young valley residents and renewed the older generation's reliance on corn. Don Modesto's story, like many others I heard, points to the aging of corn farmers and to how older residents rely on maize even though it is often not a profitable crop. Together maize production and off-farm employment constitute part of a local strategy between generations to maintain or advance the economic position of their families. But this is a paradoxical strategy: it reflects both peasants' resilience and the displacement of a younger generation. It is a strategy which also embodies strains and tensions, not only between wives and their migrant husbands but between generations, as children are raised with absent fathers and teen workers sometimes reduce their financial contributions to their parents' households.

The younger generation has joined wage labor circuits that take them further afield than previous generations. Most young men (from their teens into their thirties) now migrate to the United States for work. In comparison, young women of the same age tend to travel much shorter distances to work in maquilas and agribusiness plants within the valley, but they now do so unaccompanied by male relatives. These are *gendered labor circuits*—even though men also work in valley maquilas and to a lesser extent women become migrants bound for the United States—because they are underpinned by assumptions about what is appropriate work for each gender, and because the distances that men and women travel to work differ. These are also *transnational circuits*: migrants remit money home and live transborder lives, while workers in valley assembly plants generate profit for domestic and foreign consortiums. Participation in these circuits of labor and capital provides much-needed income for rural households, but it also puts stress and strain on families.

When economic crisis hit in 1982 San José was embroiled in a violent conflict over spring water and suffering from diminishing water levels. Local agricultural production dropped dramatically. The combined factors of conflict, agricultural problems (like pests, salinization, and insufficient rainfall, ravine, or spring water), and high inflation pushed male (and sometimes female) Sanjosepeños to search for employment in urban Mexico and the United States. While up to three dozen male residents had been employed as agricultural workers during the Bracero Program (1942–64), migration for the United States did not expand until the 1980s, when largely illegal border crossings increased in number. The first migrants in the 1980s sought agricultural employment in California, but after encountering a saturated labor market they began to find nonagricultural employment in other regions, notably in Nevada and Oregon.

In the mid-1990s migration to the United States intensified and the valley experienced a maquila boom, competing with Torreón in the northern state of Coahuila to become the "blue jeans capital" of the world (Barrios Hernández and Santiago Hernández 2004). The growing number of young valley residents undertaking migration and working in maquilas did not mark the end of the valley peasantry or its exit from the countryside, at least not yet; but it did reflect an intergenerational strategy for surmounting the difficulties of working the land and the desire of young people to go out and earn a living. As David Barkin suggests about Latin America more generally, "The peasantry is alive but not well."[1] On the one hand, maize remains the most cultivated crop in the valley and residents express a strong preference for eating criollo corn from the valley over yellow corn from elsewhere. Residents in their late thirties or older continue to identify as campesinos and to rely on maize production as a social safety net or form of risk reduction.

On the other hand, the neoliberal corn regime has accelerated accumulation by dispossession. A growing number of campesinos can no longer maintain their households by relying on the combination of small-scale agriculture, petty commodity production, and casual regional employment. Agricultural production overall is on the decline, and a number of households no longer grow maize at all, since after the implementation of NAFTA cultivating criollo maize came to cost at least as much as purchasing imported corn in the valley.

By looking at a handful of residents' stories, this chapter will highlight five main ways that livelihoods are being remade in the southern Tehuacán Valley. The first, already mentioned, is the household strategy which combines maize agriculture with migration to the United States and waged employment in maquilas. This intensified "pluri-activity," or diversification of income-generating activities, is found in many rural areas of the world but has certain peculiarities in the southern valley, such as the cultivation of commercial elote.

A second trend is the unequal sharing of the economic benefits of migration. Class differences grew in the twentieth century, before migration to the United States increased, with the formation of water societies, but these days social and economic differentiation in town is related to the distribution among households of off-farm income, particularly migrant remittances. Those with ongoing access to remittances from the United States—where wages are considerably higher than in the valley—are able to build houses, buy more expensive goods like trucks, and sometimes purchase water shares and other agricultural inputs, and start small businesses.

Other trends include the monetization of some agricultural labor. Maize was once a form of payment for agricultural labor or part of sharecropping arrangements that linked residents and households of unequal resources and status, but young men hired to work the milpa now prefer to be paid regular wages. The experience of working in maquilas and in the United States helps to discipline rural labor in particular ways; residents adapt to the rhythm and expectations of work in the maquila or restaurant and to its regular paycheck. I also noticed the emergence of a "commodification effect," as some residents evaluated their unpaid labor in the household according to wages that could be earned for comparable hours outside the household (Lem 1988; Lem 1999, 110).

A fourth trend is the aging of maize producers. Although maize agriculture was a male enterprise throughout the twentieth century —in contrast to garlic and tomato production, in which women also worked—in recent years, as young men and women have sought work elsewhere, corn agriculture has become the domain of older men. Maize production is increasingly eschewed by the younger generation. As we shall see in chapter 5, younger migrants prefer nonagricultural work when back in San José and know little about maize agriculture or seed varieties.

Finally, a fifth trend is the waning use of the intercropped milpa and of certain maize varieties. Over the past two decades one local variety has fallen into disuse, and two other criollos have become used infrequently. Together these trends provide a cautionary tale about the future of landrace diversity and abundance.

VALLEY AGRICULTURE IN CONTEXT

The recourse to corn is the peasant's last line of defense in Mexico, as suggested by Arturo Warman, quoted at the beginning of the chapter. But why would residents take up elote corn, when it is dependent on irrigation water and therefore a more costly crop than rain-fed corn? The crop's popularity was facilitated by the construction of better roads linking producers to markets, irregular rainfall, and the flexibility that the crop offers to certain producers. The valley also has an important advantage in the cultivation of elote, because other elote-growing regions suffer from frosts (see tables 3 and 4).[2]

In the mid-1970s a road was built connecting San José to its neighboring town of elote producers and buyers, San Sebastián Zinacatepec. A few years later a large market (Central de Abasto de Iztapalapa) opened in Mexico City that was a central destination for valley elote. Since most producers do not have the means to transport the crop to the market themselves, wholesalers and truckers buy elote from the farmers and take the fresh corn to regional markets as well as the cities of Orizaba, Tepeaca, Veracruz, and Mexico City. The largest profits in the elote business are made by super-

State	Area Harvested (in hectares)		Volume (in tons)		Yield (in tons per hectare)	
	Irrigated	Rain-fed	Irrigated	Rain-fed	Irrigated	Rain-fed
Baja California	887	0	4,410	0	4.972	—
Colima	519	0	4,113	0	7.925	—
Guanajuato	732	0	9,227	0	12.605	—
Jalisco	828	0	8,530	0	10.302	—
México	478	332	2,537	3,210	5.308	9.669
Michoacán	1,500	0	12,493	0	8.329	—
Puebla	10,638	0	101,836	0	9.573	—

Source: Adapted from SARH, *Anuario estadístico de la producción agrícola de los estados mexicanos, 1990*, vol. I (1992), 118.

markets, individual retailers, and food vendors who cook and sell elote with mayonnaise, chile, and grated cheese (Olivares Muñoz 1995). Before exploring the recourse to corn in more detail, some background information on maize production is required.

Crisis and Restructuring in the Southern Valley

To prepare the path for NAFTA the Mexican government restructured its agrarian reform policies. This restructuring enabled the ejido to be privately owned, rented, or sold. Additionally, a new agricultural subsidy, Procampo, was introduced to help buffer the transition to an open market. As in other regions of Mexico, economic crisis was a greater incentive than ejido reform for rural outmigration from San José (Cornelius and Myhre 1998, 14–15). Despite these reforms, which have allowed for the privatization of land grants, there has not been a rush to sell or rent land in town, particularly not to outsiders. In fact both ejidatarios and *comuneros* in San José have turned down offers from regional agribusiness to purchase tracts of land.

To facilitate the privatization of land grants a land-titling program, PROCEDE, was implemented. In San José the government began to survey communal land in 2000 as part of PROCEDE. A land survey is the first step toward registering individual land titles. After the communal or *ejidal* committee gives majority approval to

TABLE 4: Largest Producers of Elote by State, 2000

State	Area Harvested (in hectares)		Volume (in tons)		Yield (in tons per hectare)	
	Irrigated	Rain-fed	Irrigated	Rain-fed	Irrigated	Rain-fed
Colima	1,685	0	27,868	0	16.544	—
Guanajuato	1,186	0	20,701	0	17.454	—
Jalisco	1,693	426	28,670	5,020	16.934	11.784
Michoacán	1,412	0	20,940	0	14.826	—
Puebla	11,219	0	82,624	0	7.365	—
San Luis Potosí	7,421	0	87,000	0	11.723	—
Sonora	1,095	0	14,802	0	13.518	—

Source: Adapted from SAGARPA, *Anuario estadístico de la producción agrícola* (2000), 308–9.

the survey results, individuals may register their plots with the National Agrarian Registry (RAN). This enables individuals to sell or lease their plots, and to negotiate contracts with agribusiness. Thus far the communal committee has not permitted the sale of communal land, only land transfer to other locals through inheritance.

Several comuneros have expressed dissatisfaction with this process and a belief that they were losing land through the official calculation of boundaries. PROCEDE makes it more difficult for residents to inflate the amount of land in use on subsidy request forms—a practice that was popular with the introduction of the Procampo subsidy in 1993 because the subsidy was not tied to a particular crop but rather paid on the basis of cultivated hectares. The comuneros expressed their displeasure with the program by telling me, "the government is really screwing us over" (group interview, 23 October 2001).

Procampo reverses the logic of previous indirect agricultural subsidies (such as subsidized crop prices and input costs) by giving grants directly to individual producers. The idea is that Procampo should reach those who did not benefit from price supports in the past. All producers are supposed to benefit from Procampo regardless of the size of their capital or landholdings, but overall better-off farmers have benefited disproportionately (Otero 1999, 52). Sanjosepeños complain that they have waited in line for Procampo at the regional Ministry of Agriculture office alongside large landowners who did not need such government help.

As in many regions of rural Mexico, Procampo and other government programs act as a kind of rural welfare in the valley. On the one hand there is "a gradually shrinking group of viable producers who can play successfully within the new market-led rules and remain eligible for official and private lending; on the other, there is a growing group of 'the poor,' at best eligible only to receive help from government-assistance programs, which cannot help much in production" (Otero 1999, 158; de Grammont 1996).

Procampo was also designed to eliminate the intermediary between the peasant and the state by requiring each individual producer to apply for funds, in contrast to previous funding requests that went through the ejidal or communal committee. In an attempt to promote individual responsibility and discourage clientelism and corruption, Procampo is distributed through the Ministry of Agriculture (SAGARPA) rather than the PRI-affiliated National Confederation of Peasants (CNC). The rhetoric of programs like Procampo (and PRONASOL) was the reconstitution of civil society—ending paternalism, encouraging participation, empowerment, democratization —but in places like the southern Tehuacán Valley, the implementation of such programs often reproduces clientelism and corruption (Aitken 1996; Fox 1992; Nuijten 2003). Although Vicente Fox waged a much-publicized campaign to stop bribery among state officials, declaring a "war against corruption" during his presidency from 2000 to 2006, some Ministry of Agriculture officials in the Tehuacán area still expected small bribes for processing paperwork. Residents also complained that only those who are loyal to the party in office at the local level are advised when to apply for Procampo or are rewarded with state-financed projects.

While state reforms have introduced new agricultural subsidies to replace others and to buffer the abrupt transition to free trade, Sanjosepeños say they receive little help or relief from the government. Moreover, if a farmer receives Procampo it tends to be insufficient to make up for the effect of inflation and cuts to price supports and other rural subsidies (see also Myhre 1998; Nadal 2000b). Sanjosepeños report increased difficulty in affording agricultural inputs because of inflation and decreases in crop productivity. This has been so even for those few of my interviewees who had previously received support from the Banrural or other discontinued government programs. Ironically, government help arrived in San José rather late, after the economic crisis and regional water shortages

TABLE 5: Maize Production in Municipio of San José Miahuatlán, Puebla (in Hectares)

	Planted			Harvested		
	Rain-fed	Irrigated	Total	Rain-fed	Irrigated	Total
1968–69 (1)	n/a	n/a	1,340.9	n/a	n/a	1,314.2
1970 (1)	n/a	n/a	n/a	314.7	940	1,254.7
1985 (2)	121	50	171	108	43	151
1990 (3)	1,692	1,035	2,727	325	1,035	1,360
1997–98 (4)	676	655	1,331	676	655	1,331
1999–2000 (5)	1,200	850	2,050	1,200	850	2,050
2002–3 (6)	350	860	1,210	350	860	1,210
2007 (7)	350	845	1,195	238	845	1,083

Sources: (1) DGE, *Censo agrícola-ganadero y ejidal 1970 Puebla* (1975), cuadro 7, 8; (2) INEGI, *Anuario estadístico del estado de Puebla* (1990), 210; (3) INEGI, *Anuario estadístico del estado de Puebla* (1991), 186; (4) INEGI, *Anuario estadístico del estado de Puebla* (1999), 504; (5) INEGI, *Anuario estadístico del estado de Puebla* (2001), vol. 2, 595; (6) SIAP, *Anuario estadístico del estado de Puebla* (2002–3), 793; (7) SIAP, *Anuario estadístico de la producción agrícola, estado de Puebla*, online at www.siap.gob.mx (accessed 2 March 2009).

had already hit. This help was not enough to slow the damage caused by these crises, nor the neoliberal corn regime and its dependency on cheap corn imports.

Campesinos contrast the past, when water and rain were abundant and wheat, sugarcane, and tomatoes were the main commercial crops, to the irrigation-dependent elote of more recent years. After the Atzompa drought, low rainfall, and crop disease, the 1980s saw a dramatic drop in local agricultural production. Because of shortages of corn, residents had to buy yellow corn from CONASUPO government stores—which they generally use as animal feed—to make tortillas.[3] By the 1990s local production returned to previous levels and on a couple of occasions surpassed them. Early in the new century residents were reporting an overall decline in maize production, while official statistics suggest a slight drop.[4]

Despite the importance of irrigated elote in town, residents worry about irregular rainfall and their increased dependency on ever-scarcer irrigation water. "We had more corn back then because it was rain-fed," an older resident explained. Or, as another resident described the town to me, "There is no water here, there is no corn.

TABLE 6: Elote Production in the Municipio of San José Miahuatlán

	Area Harvested (in hectares)	Volume (in tons)	Yield (in tons per hectare)
1987	360	3,151	8.75
fall–winter 1989	535	4,510	8.43
spring–summer 1990	442	2,422	5.48
fall–winter 1993–94	734	5,872	8
spring–summer 1995	639	5,112	8
fall–winter 1998–99	740	4,440	6
fall–winter 1999–2000	310	1,860	6
2004	509	4,072	8
2007	514	4,112	8

Source: Adapted from F. Olivares Muñoz 1995; Anuario estadístico del estado de Puebla 1990, 230, 242, 252; Anuario estadístico del estado de Puebla 2001, 595, 636; SIAP, Anuario estadístico de la producción agrícola, estado de Puebla 2004; SIAP, Anuario estadístico de la producción agrícola, estado de Puebla 2007, online at www.siap.gob.mx (accessed 2 March 2009).

We are corn poor" (interview, 27 June 2002). One resident told me that dwindling precipitation was a form of divine punishment— recalling the local explanations of the Atzompa drought, which was seen simultaneously as the result of water overuse in the valley and as divine punishment for local violence.

San José remains an agricultural town, but there are completely nonagricultural households, dependent on the purchase of grain for making tortillas. The households that have given up agriculture altogether have done so within one or two generations. Over the course of my fieldwork in San José I encountered several households in which neither the adult son nor his father was a farmer. I met only three families that had given up maize production more than two generations back. Conversely, I also met a few former Sanjosepeño residents who commute from Tehuacán, now just an hour and a half away on the bus, to grow elote in San José.

Corn, Gender, and the Household

Corn is central to the functioning of the rural household: it is not only a mainstay of the diet and is used for sale, but its production is a coming-of-age agricultural lesson for boys and its processing in the form of tortillas is part of the expected training for girls. In addition, maize is typically at the center of sharecropping arrangements. For the first half of the twentieth century maize was also a form of payment for agricultural labor, and sometimes even non-agricultural labor, as explained to me by one octogenarian resident who as a young woman repaired and washed clothing in exchange for corn.

Men and women are responsible for different stages and aspects of cultivating the crop. In San José the milpa is a male domain. Men do not undertake the task of preparing and processing corn for food, which is considered a woman's task, while women do not cultivate corn because doing so is considered physically taxing and, when nighttime irrigation is required, too dangerous. Occasionally women will work as *destapadores* who remove excess soil covering the small corn plant after a plowing. A widow, or in more recent years the wife of an absent migrant, may have a cornfield, but the laborers (*mozos*) they hire to work it are men.

A Sanjosepeño teenager who is still in school, Frederico, learned to grow corn with his father at the age of seven, and explained to me that although women do not work in cornfields, they sometimes cultivate other crops considered less physically taxing, like tomatoes: "It's obligatory [for boys] to work in the fields, and for girls their destiny is to be housewives" (3 June 2005). Working the milpa is part of learning to be a man in San José, as well as an important part of the rural Mexican household and its unremunerated labor (Warman 1980 [1976]; Macip 1997). Yet the need for off-farm income and wage labor experience, especially since the 1980s, is altering previous configurations of gendered household labor (see chapter 5).

Residents list women's responsibilities in the household as food preparation for the family, caring for children, the elderly, and the ill, and tending turkeys and possibly pigs—but the responsibility for goats and bulls is a male purview. Food preparation includes shelling corn kernels during the corn harvest and selecting and storing seed for the next crop cycle, as well as the lengthy task of transforming corn into *masa* (dough) and then patting it out by hand for making tortillas. During special occasions like Holy Week or a wedding, women spend days preparing turkey mole and tamales. This unpaid domestic work of preparing food for the family, caring for children, washing clothes, and tending animals may be shared or allocated between women of kin-related houses. In this way the boundaries of the household, as a unit of production and consumption, do not necessarily correspond to an actual, physical house.

Gendered work is mobilized through discourses about the patriarchal family, which are based on assumptions about the husband as the decision maker and head of the household. In reality, though, important decisions are not always unilaterally decided by the husband, and may be worked out between spouses. And while the husband/father is considered the main provider of maize and income for the household, the different members of the household, young and old, male and female, tend to contribute in some way to the household's supply of corn and income.

Sanjosepeños discuss women's domestic work as a duty to the family rather than "work" (trabajo), because it is not as physical as the work in the milpa, nor does it bring in income. This is so even though girls and women are engaged in various income-generating activities from home, including piecework embroidery, subcontracted apparel work, and the operation of small taco stands, corn mills, and corner stores.

Maize Agriculture

Although the Tehuacán Valley is lucky not to suffer from frosts, agricultural producers face the problems of salinization of the soil and the formation of hardpan below the surface. Salinization is the buildup of calcium salts and carbonates in the soil, which can be toxic to plants. The concentration of calcium can also reach such high levels as to form a hardpan, or hard layer, beneath the surface of the soil, making proper drainage extremely difficult. The climate of the valley, which is semi-arid, contributes to this problem, as evapotranspiration exceeds rainfall (Enge and Whiteford 1989, 28). The rainy season usually occurs in May and June. After a couple of months of lowered rates of precipitation, rainfall increases again in August and September before dropping off considerably (ibid., 25). The mean temperature ranges from a low of 12 degrees Celsius to a high of 35, with the hottest temperatures in April through June. In the hot months temperatures can reach 40 degrees. San José is in the southern end of the valley, bordering Oaxaca, and its climate is often a few degrees warmer than in Tehuacán.

For those agriculturalists with irrigation water, the first planting cycle in San José begins in January or February, and crops are harvested in the summer. For most residents maize is planted during the rainy season, in May or June, with the harvest in the fall. Some residents plant a second milpa in the fall following the harvest, which is ready the following February or March. There is some variation in the planting cycle for agriculturalists, depending on when their irrigation water is distributed, the amount and timing of rainfall, and the location of their land.

The type of crop chosen by farmers also depends on whether their land is close to an irrigation canal or the ravine (which fills with run-off water in the rainy season) or on rain-fed land.[5] Campesinos prepare their land by removing rocks and shrubs and plowing (barbecho) the land with a rented machine or a steel plow pulled by oxen (yunta). Then an irrigation soaks the soil to reduce its saline and alkaline levels. A week later the campesinos harrow or comb the soil to evenly distribute it (rastrear). This is followed the next week by a furrowing of the fields (surcado) and the immediate planting of seed (sembrar), which follows behind the furrow plow. Corn kernels are dropped into a hole shoveled into the furrow. Where the corn is part of a traditional milpa, beans are tossed into the same hole—although this has been a less common practice in recent years. Generally campesinos use gallinaza (chicken droppings mixed with other fertilizer) as their fertilizer, which is produced by the large poultry industry in the valley.

The fields are irrigated ten to fourteen days after planting, which is about a month after the first soaking of the soil. When the water arrives it is forced into the field by opening a small dam from the feeder canal. The third, fourth, and fifth irrigations take place 45, 75, and 105 days or so after planting.[6] The last round of irrigation is optional and depends on whether the maize is to be used as grain for the household (since harvesting the crop as elote requires one round fewer of irrigation), and whether a second crop will be planted soon after harvest. Repeated irrigations and high winds sometimes wash away the soil around the base of the plant, making it more vulnerable to toppling over. Cultivators may shore up the base by passing by on a plow pulled by bulls or oxen and tossing dirt against the plants. To make sure that the corn plant (which is thirty or forty centimeters tall) is not completely covered by the soil, campesinos, sharecroppers, or hired laborers (mozos) go by and uncover the plant by hand (destapar). If elote is sold, the purchasers provide the labor for the harvest and truck the corn out of the valley. Usually

Young boy standing over drying corn

a small portion, up to 5 percent, is left for use as seed. If the corn is not sold it is left to dry in the field, harvested, and then taken home to be dried out in the sun on mats.

THE RECOURSE TO CORN

As in other parts of Mexico, maize is an important crop in the valley because it is the cornerstone of the diet, part of the habitus of everyday life, and because it can be diverted into exchange in times of crisis or income shortfalls. Households may grow rain-fed corn with the intention of consuming it in the form of tortillas. Along the way the corn may be "diverted" into the commodity phase and sold for money (Appadurai 1986, 26), usually at a loss. Since rainfall is irregular in the valley, subsistence rain-fed corn sometimes requires a round of irrigation.

In contrast, elote (a variety called *chicuase*) is grown for sale but is sometimes withdrawn from its commodity phase, dried as grain, and consumed in the producer's household in the form of tortillas. Irrigated elote is a particularly flexible variety of corn because it can be sold on the market for a better price than grain or dried and consumed at home. But because elote is more costly to produce than rain-fed corn, the poorer strata of peasant households either share-crop elote fields or grow only rain-fed corn.

Cost

Elote production is undertaken by those households with sufficient economic or social resources. Farmers require water shares or a source of income like migrant remittances or the Procampo subsidy to purchase their inputs (such as fertilizer and tractor rental) and, more commonly these days, hired labor in the field. If a farmer without such resources is respected as reliable by neighbors and relatives, including fictive kin, he will team up with others and work the field for a share of the crop. The practice of sharecropping enables resource-poor peasants, particularly those without water shares, to grow elote. And for those campesinos with water or an income, sharecropping provides labor for the milpa that is paid with part of the crop rather than cash.

Elote generates a small income for farmers if they are water socios, they have a good harvest, and the harvest coincides with a high market price. If these factors are not in their favor the sale of elote does not cover the cost of production. When I began my fieldwork in the valley the cost of growing a hectare of irrigated elote was $657 U.S. (see table 8). There were many variables affecting the production costs and the level of profit, such as the amount of rainfall, the cost of irrigation water, the availability of labor, and so on. If the farmers had a good yield and their crop was ready for sale when the price was high, they made a profit of around $828 after paying for irrigation water and labor. If their crop produced a low yield, their profits would have been $271 (see table 9).[7] It should be pointed out, however, that elote farmers generally keep some of their maize for household consumption, thus reducing the level of profit. When there is not enough rainfall, growing rain-fed maize (nahuitzi) can cost about the same as purchasing criollo corn in the form of grain. It is also substantially less expensive than purchasing pre-made tortillas for household consumption.

A young family of five requires a minimum of one hectare of rain-fed maize to make enough tortillas for a year, but it would still have to purchase some local grain when its own supply of corn ran out.[8] During my fieldwork the cost of producing a hectare of rain-fed corn ($282 or more) plus the cost of a few months of purchased grain to make up for any shortfall ($65) was about $347. If the family needs to hire labor to harvest the corn, or its field requires a couple of rounds of irrigation, its production costs are higher (see table 10).

If the same young family of five did not grow corn but purchased local grain, the cost would be a minimum of $546 per year[9]—less if the family bought yellow corn from elsewhere. Sanjosepeños frequently told me that growing maize cost at least as much as purchasing it.

In the countryside adults eat anywhere from six to twenty tortillas daily, depending on the size of the tortilla, which varies from region to region.[10] Buying fresh tortillas is the most costly method of acquiring the household's daily staple, and for this reason most women prepare tortillas themselves, only occasionally purchasing pre-made tortillas from their neighbors. Angela, whose house has one adult and four children (see below), takes one hour to prepare the tortillas to feed her family. When the children were young they consumed two kilos of tortillas every day, with the leftovers going to feed the turkeys. If the family had bought tortillas for that entire year, that would have cost around $750. Four years later, when three of the children were teenagers, the family nearly doubled its consumption of tortillas. If Angela had bought 4 kilos of tortillas instead of making them at home, it would have cost her about $1,805 a year.

Cost is clearly one consideration in a household's decision to grow corn and the elote variety. Yet residents repeatedly say in daily conversation that corn, including elote, is not a profitable crop. The most common reasons residents gave for why people continue to grow corn had to do with taste and risk, not profit or weighing the costs.

TABLE 7: Costs of Elote Production, Spring–Summer 1995

Activities	Amount by Hectare	Unit Cost	Pesos per Hectare
Preparing the Land			
Barbecho (to plow)	5 hours	30	150
Rastra (harrow)	3 hours	30	90
Surcado (furrow)	3.5 hours	30	105
			345
Planting			
Seed	4 boxes (cajones)	12	48
Labor	4 day laborers	20	80
			128
Fertilization			
Fertlizer (urea)	4 packages	60	240
Fertilizer (gallinaza)	8 tons	80	640
Application	2 day laborers	20	40
			920
Labor			
First labor	3 yokes	50	150
Remove soil, uncover plants (destapadores)	2 day laborers	20	40
			190
Irrigation			
Irrigation water	5 irrigations	160*	800
Application	2 day laborers	20	40
			840
Harvest	covered by purchaser		
Total			2,423 (U.S. $315. 87)

* The cost of irrigation relates to the flow and volume of water in the canals. The quantity here is 25 liters per second.

Source: F. Olivares Muñoz, 1995. Translated by author. U.S. dollar amounts based on exchange rate as of December 1995 ($1 = 7.6708 pesos).

TABLE 8: Costs of Elote Production, Spring–Summer 2001 (Rainy Season)

Activities	Amount by Hectare	Unit Cost	Pesos per Hectare
Preparing the land			
Remove rocks, shrubs	2 day laborers	80	160
Barbecho (to plow)	6–8 hours	100	800
Rastra (harrow)	6 hours	100	600
Planting			
Seed	5 boxes	15–18	90
Labor	4 day laborers	100	400
Fertilization			
Agrochemicals	3.5 bultos	109	381.50
Fertilizer (*gallinaza*)	8 tons		1,300
Labor	2 day laborers	80	160
Other Labor			
Surcado (furrow with yokes)	2	160	320
Destapadores (uncover plants)	2	80	160
Irrigation			
Irrigation water	5 hours × 5 irrigations	80/hour*	2,000
Harvest	covered by purchaser		
Total			6,371.50 (U.S. $657.28)

Total if received Procampo:** 6,371.50 − 829 = 5,542.50 (U.S. $571.76)
Total if seed, water, and labor (planting and fertilization) owned by producer: 6,371.50 −
   2,650 = 3,721.50
   (U.S. $383.91)
Total if seed, water, and labor owned by producer and received Procampo: 6,371.50 − 3,479
   = 2,892.50 (U.S. $298.39)

*This is the minimum amount for irrigation water. The price of irrigation water can reach 110 pesos per
   hour during the fall–winter season.
**Procampo paid 829 pesos per hectare in 2001.
Source: Data collected by author.

TABLE 9: Elote Profits, 2001

Yield in Bultos (bags) per Hectare	High Purchase Price Total Paid	Total Profit if Water Bought (in Mexican Pesos)	Total Profit if a Water *Socio* with Labor and Seed
80	180 pesos per bulto 14,400	14,400 − 6371.50 = 8,028.50 (U.S. $828.22)	10,678.50 (U.S. $1,101.60)
50	180 pesos per bulto 9,000	9,000 − 6,371.50 = 2,628.50 (U.S. $271.16)	5,728.50 (U.S. $590.95)

Taste

Most residents prefer local or regional criollo corn for making tortillas, even though during my fieldwork imported or industrial grain
was 40 percent cheaper at the local market. Ángela, for instance,
who has a small milpa for some of her household needs, prefers
tortillas made from the white, local criollo corn and pays extra for it
at the market when her own supply runs out.[11] A female storeowner
who also has a small milpa for her household consumption told me
that people grow corn despite the cost because they prefer the taste
of white corn and it makes tortillas of better quality than those
found in Mexico City, which are made from yellow corn or industrial
corn flour. Yellow corn is usually fed to farm animals: "We grow
corn because we want to have good, soft, white tortillas. They don't
turn out the same in the city. In Mexico City, a truck carrying *masa*
[dough] comes round as if it were mud. It's even uncovered! They
say we live like animals here in the countryside, but in the city, they
eat like animals!" (interview with female storeowner, 20 June 2006).
When yellow corn is received for free through government programs, or purchased at the local store, it is either fed to farm animals in San José or mixed with local corn to hide its taste and
texture. With her comparison between rural and urban tortillas, the
storeowner counters urban stereotypes about rural Mexico as backward or uncivilized by suggesting that access to local corn of high
quality is more dignified than the "animal feed" that urban Mexicans eat. She conveys the sense that good corn is part of a dignified
life, a sentiment which I explore further in chapter 5.

TABLE 10: Costs of Maize (Grain) Production, 2001

Activities	Amount by Hectare	Unit Cost	Pesos per Hectare
Preparing the Land			
Barbecho (to plow)	6–8 hours	100	800
Rastra (harrow)	6 hours	100	600
Fertilization			
Agrochemicals	3.5 bultos	109	381.50
Fertilizer (*gallinaza*)	8 tons		1,300
Labor			
Surcado (furrow with yokes)	2	160	320
Destapadores (unblock canals, deweed)	2	80	160
Total			3,561.50* (U.S. $367.40)
Total Cost if received Procampo	829 pesos		2732.50 (U.S. $281.88)
With irrigation**	5 hours (× 2 days)	80/hour	+800 (U.S. $82.53)
Total with irrigation and Procampo			3532.50 ($U.S. 364.41)

* Two producers reported that the production costs of rain-fed maize were as low as 2,000 pesos a hectare. Various factors affect the overall cost, including access to unpaid labor, ravine water, rainfall, etc. Note that the total cost with two rounds of irrigation and no Procampo would be up to $449.93.
**Some farmers report buying up to five rounds of irrigation. This would bring their production costs to $574, more than the minimum cost of local grain needed to feed a young family of five ($546).
Source: Data collected by author in 2001. U.S. dollar amounts based on exchange rate as of February 2001 (1 peso = $0.10316).

As a Safety Net

Valley residents grow elote and rain-fed maize for consumption either because doing so is cheaper (because they are water shareholders or socios with good land and sufficient labor) or because they consider it less risky to grow than to buy. For residents without access to water or unpaid labor there is little price difference between growing maize and purchasing local or regional grain, yet they may continue to grow corn because they feel it is the more reliable option. For instance, one mother in her thirties who hires labor for her milpa does not rely on the market alone for her grain. She explained her approach with this question: "What would happen if there was no corn for sale?" (18 November 2002). When I asked one maize farmer in his late sixties why people grow corn when doing so is not profitable, he answered: "There is no work here except in the fields. If you are older, they're not going to give you a job. I went to find out about getting contract work in the United States, but since I'm sixty-eight years old, they told me 'no'" (25 June 2006). Since older residents are unemployable, their milpas are all they have to help maintain the household. Rain-fed maize and irrigated elote afford older valley residents a degree of stability in weathering the impact of fluctuations in income and inflation, as well as market shortages of grain. Maize production provides security to rural households in the absence of unemployment insurance, pensions, or a state-sponsored social safety net, particularly for older residents. But at the same time, corn agriculture (or agriculture more generally) is not a viable livelihood on its own. A fifty-year-old resident who had returned to maize agriculture after inheriting water shares from a parent and working for more than twenty years as a truck driver, explained that no one (or no one household) could rely on corn production alone: "I work the milpa but if we only rely on the fields, after five months, where are you going to get your food? You can't live off agriculture alone. When your corn runs out, what then?" When asked what he thought would happen in the future he replied, "In the future, no one will grow maize. Everyone will have to make their living elsewhere, except for a few who will continue to grow so people can buy corn locally" (interview, 22 June 2006).

Residents in their late thirties or older rely on the recourse to corn; they tend to view the cultivation of maize as minimizing risk. In contrast, migrants and maquiladora workers in their teens and twenties do not feel that maize agriculture provides such advantages, particularly not for their generation. Migrants in their thirties can be seen as semi-proletarians based on their experience of going back and forth between household agriculture and work as a migrant in the United States, whereas younger migrants and maquila workers are fully proletarianized yet flexible workers. As disposable and casual workers, whose labor experience and preference lie beyond the farm, they are dependent on the labor market.

Throughout the twentieth century Sanjosepeños traveled to sell their goods or earn a wage as cane cutters or domestic workers, but their journeys were not as long as they are today, and theirs was not a transnational town. These days Sanjosepeño households combine subsistence agriculture and petty commodity production—in this case elote production—with income from other sources, including construction, agricultural day labor, transnational and national migration, small businesses (such as bakeries, small stores, and corn mills), piecework embroidery, work in maquilas and agribusiness plants (such as hatcheries), and selling goods at the valley markets (*tianguis*).

During the 1960s entrepreneurial mestizos from Tehuacán set up capital-intensive farms (*granjas*) in the valley, producing poultry, eggs, meat, and later animal feed and bird vaccines for export. Led by the Romero family, well-known producers of eggs and poultry, this industry bought land in the valley and obtained pumping rights at a time when the further expansion of irrigation pumps and galerías was prohibited (Enge and Whiteford 1989, 179). This drove up the price of land and water in the valley, putting these resources further out of reach for smallholding campesinos (ibid.). During this period the city continued to be an important maker of bottled water and soft drinks, and from the 1970s apparel factories made uniforms for the domestic market. In the 1990s the industry expanded as Mexican apparel companies linked up with foreign consortiums.

Maquilatitlán

With NAFTA the region experienced a maquila boom. Manufacturers of blue jeans and apparel for export quickly set up maquilas in both the city and the valley. They became the largest source of employment in the city, with seventy thousand employees in over seven hundred registered and clandestine factories. At its height the industry produced fifty million articles of clothing per month, largely for the Mexican and United States markets under brand names like Guess?, Levi's, Wrangler, and Tommy Hilfiger (Barrios Hernández and Santiago Hernández 2004, 25, 30). Attracting young, indigenous men and women from rural areas of the region and neighboring states to work in its maquilas, the city was nicknamed the "capital of blue jeans" and "Maquilatitlán"—which combines the Nahuatl suffix for "place near an abundance of" and the Spanish word maquila. President Vicente Fox's administration promoted the maquila industry of Tehuacán as a model for the development project Plan Puebla-Panamá, which focused on fostering agro-industrial and maquila production in southeastern Mexico (ibid.).

Children as young as nine work in maquilas, and while both men and women are employed in the industry, their jobs and salaries differ; men are hired as cutters and in laundries, where they are paid higher wages, while women are sewers. In 2002 the average salary for sewing work was between $38 and $82 per week, while those working in the industrial washers made up to $130 (ibid., 49). Meanwhile the estimated household cost for a family of five was $106 per week (ibid. 51). Young women from San José view maquila employment as a necessary source of income, but this work also tests the limits of what their neighbors and relatives consider appropriate behavior for women (see chapter 5).

The manufacturers assemble imported precut garments or follow the "full package" system, which coordinates the different stages of production from the acquisition and cutting of textiles to assembly of the garment, finishing, packaging, and in some cases distribution of the finished product (ibid.). The product is then exported—with 20 percent sold in Mexico—while industry pays tariff only on the value added. Mexico offers exporting companies simplified and expedited customs and regulations, subsidized infrastructure, tax exemptions, and a low-paid, disposable workforce. The maquiladora system is such that what "Mexico essentially exports is its labour force, without ever having to leave the country" (Delgado Wise 2003, 6, citing Carlos Tello).

The work in maquilas is typically on an assembly line. Each zone in the factory works on a particular stage in making a pair of blue jeans or apparel. As a piece of clothing passes through the various stages of production, workers stationed in their zones repeat the same process throughout the day. This repetitive work takes place in factories which are hot and loud, with vibrations from industrial sewing machines.

In their study of the Tehuacán maquila industry, two human rights and labor organizers, Martín Barrios Hernández and Rodrigo Santiago Hernández (2004), found that some companies failed to register their workers for the government social security program (IMSS), that workers were docked pay for mistakes or slow work and were required to remain at work until their quotas were fulfilled, without paid overtime, and that those working in the laundries were exposed to toxic fumes. Female employees face sexual harassment and racial slurs from their superiors. The industry has opposed labor organizing, particularly the establishment of independent unions, and organizers have been threatened and beaten. One of the study's authors was himself threatened and beaten as well as arrested on trumped-up charges.[12] Some valley maquilas are affiliated with large, official unions (such as FROC-CROC, CROM, and CTM), which are linked to the government and charge workers obligatory dues without negotiating better working conditions or wages (ibid.). Some improvements to work conditions were made, however, after anti-sweatshop campaigns in the United States, Canada, and Mexico drew attention from the news media (ibid.). Maquilas also pose environmental problems for an already overburdened water and agricultural system. The runoff water from factories with industrial washers that chemically treat, wash, and dye jeans can be seen in maize, fruit, and vegetable fields that line the highway taking commuters to and from the city.[13]

In 2000 the maquila boom ended. Some twenty thousand employees lost their jobs when foreign clients closed their doors to move their operations abroad in search of even cheaper labor.[14] But as maquilas closed, the industry expanded its putting-out system, which relies on subcontracted sewing workshops and employees working from home. These are often clandestine operations. The industry incorporates different generations of women into distinct phases of the production process. Often women's work trajectories begin at a maquila before marriage and continue with work at home or in a workshop after having children (Flores Morales 2008). In other words, these are transitions related to the women's life stage and household circumstances. Valley workshops pay less than maquilas, but they provide young mothers with the ability to work around the school schedule of their children. This is particularly important for single women or those without a kin network that can take care of their children.

In addition to workshops, mothers with small children and women of an advanced age are incorporated into the production process through homework. These women work from home, sewing buttonholes or pulling out hanging threads from clothing assembled in maquilas. Homework enables women to recruit their daughters and sometimes even male relatives to help with the work. In the process young girls are trained and socialized to contribute to the household unit (ibid.). While this maquila homework is found throughout the valley, at the time of my fieldwork in San José it was more common to find women working as piecework embroiderers at home for valley entrepreneurs, largely from Chilac, who use the embroidered patterns in clothing marketed as traditionally indigenous for tourists in Oaxaca.

The making of a flexible labor force is clearly a gendered process in the valley. Young men prefer labor migration to the United States over maquila or granja work, turning to valley industry as a temporary step before migration or as their last option. In contrast, for women the maquila industry is one of few options for generating an income (Flores Morales 2008).

Transnational Migration

Mexico is the world's largest exporter of migrant labor. These migrants have been integrated into a restructured United States economy characterized by the internationalization of production and the precarization of labor, as permanent workers are increasingly replaced with temporary ones (Delgado Wise and Márquez Covarrubias 2008). During the 1990s the flow of migrants to the United States was ten times higher than that of the two previous decades. An estimated ten million people born in Mexico live in the United States with or without documents, and together with United States citizens of Mexican ancestry they constitute the world's largest diaspora located in one country (ibid.). The "migradollars" that migrants send home are the second-largest source of foreign exchange earnings in Mexico, a contribution larger than either tourism or agricultural exports. They totaled $25 billion in 2008.[15] As the first decade of the twenty-first century ended with a global financial crisis, unemployment in the United States rose and the amount of remittances sent home by Mexican workers dropped.

Three out of four agricultural workers in the United States are Mexican-born, and Mexicans increasingly work in construction, manufacturing, and the service sector (Delgado Wise and Márquez Covarrubias 2008, 1367). Unlike earlier Mexican migrants who were typically agricultural workers, more Mexicans now work in United States industry than in agriculture.[16] In recent years Canada has begun to rely more heavily on Mexicans as contracted temporary agricultural workers, but the number is small compared to the equivalent in the United States (Binford 2002).

With this growth in migration, more women and families are making the journey northward, and from areas outside the traditional sending regions of western Mexico. San José is one such area, where extensive migration networks developed in a short time (Binford 2004).

Migration from San José to the United States is largely cyclical rather than unidirectional. Ranging in age from their teens to their early forties, men and to a lesser extent women leave the valley to work or live in the United States. Their place in the life cycle and in the household influences their reasons for working in el norte, how much money they send home, and to what use they put their earnings. The youngest among them send remittances home to their parents or new spouses in the hopes of eventually establishing their own households, while those in their thirties or forties send money home for the upkeep of the household. The destinations of Sanjosepeños are overwhelmingly Oregon, California, and Nevada rather than the New York and New Jersey region, where most of the estimated one million poblanos in the United States work (INEGI 2005, 20).

These are transnational lives. Migrants are simultaneously engaged and embedded in the social life of their valley town and of their destinations across the border. Young migrants often marry other Sanjosepeños and set up households in town, while older migrants set out already married. Close to a third of "migrant" interviewees (or residents who had been in the United States within five years of the interview) had bought land in San José to build a house, were in the process of building a house, or had already built a house in town. All but two said that they imagined their future in San José. Two migrants had even returned home to become mayordomos.

San José is part of a transnational social field which links those people who stay behind to those working and living abroad through emotional bonds, financial transactions, shared household and familial responsibilities, milpa production, community celebrations (like mayordomías), and political affiliations (Glick Schiller 2004; see also Basch, Blanc Szanton, and Glick Schiller 1995; Rouse 1991;

Smith 1998). Social relations are stretched as people, goods, money, and ideas cross the border. In their everyday lives residents in San José interact with those abroad through phone calls and money transfers. But they also negotiate their absence in multiple ways, such as by managing the household milpa when there is a shortage in agricultural labor. In this way even those residents without migrant household members live in a transnational social field.

Migrant spouses and children contribute to the rural household in San José. Decisions about management of the household and how to spend earnings are negotiated between those present in San José and those abroad. As with Angela, whose story is recounted below, the discussions can cause tension between spouses. Men make up the majority of migrants from San José, but this does not mean that women are the passive recipients of migrant remittances from male relatives (Kunz 2008). Rather, women negotiate the use of the money, contribute to the household's upkeep and income, and if they are in their own home rather than the home of their mother-in-law, also manage the household.

Migrants who set out in the 1980s often went to California to work as agricultural hands. But since the mid-1990s the nature of employment has changed. Most migrants from San José work in restaurants, food-processing plants, factories, and casinos in Las Vegas, instead of agriculture. Further, what was once an eight- or nine-month work cycle abroad followed by several months at home (during the October harvest and the year-end religious holidays) can now stretch to two, three, or even five years abroad. These extended stays put extra tension and stress on separated families. When back in San José, migrants in their teens and twenties work in valley industry or just pass the time hanging out until their money dries up and they decide to make the journey north again. In effect they have little agricultural experience either at home or abroad. Older migrants, in their thirties and forties, are more likely to work the fields or in the construction business when they are back home.

In 2003 a Californian company began to recruit in San José and other valley towns for tomato workers. If such contract work expands in the future, more young men will have agricultural experience. But for the time being most migrants work in the service industry and prefer nonagricultural work. One clear advantage of tomato work in the United States is that it is contract employment and does not require dangerous and expensive border crossings.

Deportability

While the accelerated migration of the 1990s made use of networks of friends and kin already abroad, migration was not less risky or expensive than for those who had left earlier. In the mid-1990s Washington's border enforcement strategies unintentionally redirected illegal border crossers to more dangerous routes. Additionally, since September 11, 2001, migrants have reported more difficulty in crossing the border illegally and as a result return home less frequently. Migrants from San José cross the border with the help of guides called "coyotes," whose services cost between $1,500 and $2,000 (during 2001–6). Sanjosepeños arrive in the United States as undocumented migrants who are either looking for work or fortunate enough to have a job waiting.

What makes Mexican labor particularly vulnerable, temporary, and replaceable is its "illegality" and therefore deportability. Nicholas De Genova (2005) explains that "the legal production of 'illegality' [through immigration policies and the policing of the border] provides an apparatus for sustaining Mexican migrants' vulnerability and tractability—as workers—whose labor-power, because it is deportable, becomes an eminently disposable commodity" (215). The possibility of being deported is not only an important part of the everyday migrant experience but helps to keep migrant wages down. Mexican migrants are the lowest-paid workers in the United States (Delagado Wise and Márquez Covarrubias 2008, 1368). When United States immigration policy toward Mexico shifted in the mid-1960s, it created a de facto guest worker program that hinged upon transforming Mexican migrants into undocumented or illegal workers. In other words, the role of United States immigration policy and border control has been to operate "the border as a revolving door, simultaneously implicated in [worker] importation as much as (in fact, far more than) deportation" (De Genova 2005, 248).

Remittances as Development

Studies of Mexican migration tend to focus either on the social costs of migration or on its economic and social benefits. Jeffrey Cohen (2001, 955) nicely summarizes these differing approaches as dependency or development models: "Dependency models argue that migration exacerbates local socioeconomic inequalities, increases economic dependency, and drives unproductive consumption within peasant households while creating pools of cheap labor waiting to be exploited. . . . Development models emphasize the benefits of migration and the positive outcomes that are possible with the careful investment of remittances." Often home communities experience both the social costs and the benefits of migration, and do not always fit one model or the other. What Cohen calls a transnational approach to migration suggests looking at both consumption and production practices, both the social costs and the benefits of migration, and questions of both class and ethnicity in migrant communities (Cohen 2001). In chapter 5 we explore some of these issues more closely.

For researchers who focus on the stages of migration, the first stage displays increased social and economic differences in rural sending towns, while in the second stage, when migration has become a more widespread practice in the sending community because the costs and risks of migration have decreased, the standard of living improves and social inequality decreases.[17] Migration from San José can be characterized as being in its second stage, because it is a widespread practice and residents journey north on routes that they themselves or others known by them have already taken. They rely on developed networks of family and friends in the north, sometimes on both sides of the border, for a place to stay or to secure employment. During my research I found that although migration from San José to the United States is a common practice, this has not translated into fewer risks or lower fees. New migrants and many returning migrants rely on paid coyotes, and in recent years attempts to cross the border have been more likely to fail. Nor does it appear that widespread migration contributes to a lessening of social inequality in town.

In San José residents say that they are better off than previous generations because of their access to jobs in the United States. For some Sanjosepeños migration is an economic necessity, while for others it represents a choice to look for better opportunities, part of an economic advancement strategy. Several interviewees also discussed migration as a source of pride, and as a means to start their own household or to contribute to that of their parents. Residents point to the economic benefits of migration but also discuss its social costs, including the scarcity of agricultural labor and the separation of their families. They worry about exploitation on the job and the physical dangers of crossing the border illegally. Undocumented border crossing is a dangerous activity.

Leigh Binford (2003) has criticized the development approach to migration (or what he refers to as a functionalist approach) for assuming that the investment of remittances in sending communities generates long-term local employment; that there is no saturation point for local investments; and that all economic strata are participating or benefiting equally in migration. In response, Cohen, Jones, and Conway (2005) argue that remittances cannot be blamed for the underdevelopment of rural Mexico, that they provide key support to migrant households and also contribute to a decrease in class differences within communities and particularly between rural and urban Mexico. The authors contend that it is unrealistic to base the success of migration and remittances on whether they "develop local and regional economies to the point where the majority can sustain a dignified lifestyle" (Cohen, Jones, and Conway 2005, 88). In San José remittances are indeed central to the reproduction and even the improvement of rural households, but I believe that Binford's point should not be lost: transnational migration is not simply economic development. Nor is transnational migration the cause of rural underdevelopment, but rather an effect of it.

Even though migration to the United States does bring economic and social benefits to rural households, I agree with Binford that remittances should not be seen collectively as a development success story: they come at a high social cost and do not necessarily generate long-term non-migration employment alternatives. While there are migrants who use their earnings to finance small business ventures in their hometowns and migrant organizations that pool funds for projects in rural communities or use the government Tres por Uno program (a development matching fund) to channel remittances toward public projects in their hometowns (Fox and Bada 2008), most remittances in Mexico are destined for household consumption (Delgado Wise and Márquez Covarrubias 2008, 1372).

Having helped push rural Mexicans toward migration to the United States, the state's approach to development rests on the remittances sent by migrants. Indeed, migrants are often heralded as the heroes of development. In "the absence of a real development project, migrants . . . are held accountable for promoting progress in a situation where the state, claiming minimal interference, declines to take responsibility" (Delgado Wise and Márquez Covarrubias 2008,

1372). As Cristóbal Kay wrote (2008, 936) about Latin America more generally, "It is paradoxical, ironic and tragic that perhaps the greatest contribution to rural poverty reduction has come from the poor themselves, from those who have emigrated and sent remittances to their families."

Remittances in San José

The decision to stay or leave is varied, influenced by the life stage of potential migrants and that of their households. Some migrate to earn money for specific goals like building a house in San José or buying a truck, while others leave to provide ongoing support for their families. The density of the potential migrant's social network also factors into potential migrants' decisions—if they know people at the border, if their friends or relatives are making the journey, if they have a place to stay in the United States, if a friend or relative can help them secure employment, and so on. Other factors include the availability of funds to make the trip. Many residents borrow from local lenders to make the journey and pay them back with interest from their first earnings.

The needs and wants of different members of the household contribute to residents' decisions about whether and when to head north. Teen residents sometimes face pressures at home to leave school and work in the United States or work in maquilas. Two teen interviewees described this sort of pressure to me, explaining that their parents needed their income. Young residents may also migrate to help a sibling stay in school or pursue some training for future employment. Their income is related to the livelihood strategy of their hometown household.

There is a large range in the amount of money sent home, with migrant interviewees reporting that they send between $200 and $600 a month while employed. Several interviewees reported savings of over $1,000 a month. This variation can be attributed to the type of employment secured in the United States, the debts incurred while crossing the border and before finding work, the migrant's English proficiency (which can mean the difference between a job dishwashing and one waiting tables), how frugal the migrant is, and whether the migrant had their own house before migrating or was in the process of saving and building one in San José.

When young migrants remit earnings, they send it to their parents for household costs like food and consumer goods, healthcare, leisure, or building a home. As can be expected, when a migrant or young couple decides to establish a household in town, or when a young woman marries or becomes a mother, the parents receive a smaller part of the remittances, if any. Maquila workers who live with their parents also contribute earnings to their parents, while keeping some spending money for clothes, leisure, and small consumer items. Slightly older migrants, in their thirties and forties, report remitting savings for household maintenance and reproduction, a small business, or specifically for agricultural inputs or maize production. In these cases it is not only the household that is transnational, but also the milpa.

Not surprisingly, younger migrants, unless they are engaged to be married or expecting children, typically send less money home and spend more abroad. They report remitting money specifically for agriculture less frequently, and the parents of younger migrants sometimes complain about wasteful or neglectful migrant sons.

Several migrant interviewees reported using their earnings for the purchase of items beyond mere social reproduction and consumption. They purchased land, built a house, or bought machinery or a truck to start a small business or invest in agriculture. The purchase of a truck might create work for a handful of people in town if, for instance, it was used to open a construction business. But as Binford (2003) asks, how many such businesses can open in a town of nine thousand before the local market becomes saturated?

All the financially better-off households that I encountered included migrants currently abroad or with a bracero past. Better-off households are defined as those that had water association shares, at least five hectares of land, or income-generating equipment like a tractor, truck, or electric corn grinder (molino). Water association shares are important since irrigation water is one of the biggest expenses facing agricultural producers in the valley, and as I demonstrated in chapter 3, the valley is highly stratified between producers who own water resources and those who do not. But as the case studies discussed below illustrate, migration does not always raise a household's standard of living. Less well-off migrants reported not earning enough money in the United States to be able to save any money; or reported spending on household reproduction and flexible costs like a health crisis or a wedding party.

Remittances do enable some households to move up the social

and economic ladders. Don Ignacio, discussed in chapter 3, once worked as a bracero and had purchased land and water shares with his savings in the 1950s. As an early migrant, Don Ignacio gained social prestige and economic standing in the community based on his experience abroad and his purchase of water shares. José (see below) was able to buy agricultural land, a tractor, and a water share after years of working abroad, but he also came from a family that already owned water shares. José's priorities today are to invest in agricultural machinery, because he "was taught to think about the future." Migrants who purchase water shares can translate their economic advantage over non-socios into social prestige and political influence.[18]

Transnational Milpas

Although migration has meant that there is a shortage of young men available or willing to sharecrop or work the fields, migrant remittances and maquila work contribute both directly and indirectly to the production of maize. While only eight of the thirty maize farmers whom I interviewed reported direct spending of remittances on agricultural costs (see Appendix), more significant was how widespread remittances were to maintaining Sanjosepeño households. Of the seventy residents in total whom I interviewed, forty-seven, or 67 percent, reported either having migrants in the immediate household or being migrants themselves. Of this same number, forty-four, or 63 percent, had grown corn at least once in the past year.

Migrant and maquila earnings help free up other income in the household to cover the costs of maize production. With expenses like consumer and household goods covered, producers are able to sell part of a crop or an animal, or apply for Procampo and use the cash to buy agricultural inputs. When one farmer responded to my questions about using remittances he hinted at this relationship: "No, my son doesn't send money for use in *el campo*. The money he sends from the United States helps us with our other children. But when I need to buy fertilizer [for the milpa], I sell a little grain at the local market or to my neighbors" (12 February 2004).

Through a combination of outmigration and elote production, Sanjosepeños have adapted to economic crises and drought, and in some cases improved their standard of living. But this inflow of migrant remittances does not necessarily help to create employment alternatives to migration, nor help the long-term reproduction of maize agriculture.

A closer look at a few residents' stories will illustrate how the transnationalization of the household is affecting local agriculture and the social relations of production. More particularly, the un-remunerated labor of the household—the masculine work of the milpa as well as the female domain of food preparation and child-care—is either undergoing monetization or a commodification effect. A young labor force is accustomed to an hourly wage or at least thinking in terms of hourly wages. Sharecropping has declined and milpa work has increasingly become monetized. Many households pay day laborers in cash for working in elote fields and sometimes even subsistence corn production, when sons and husbands are abroad.

Although women are occasionally responsible for off-field decisions about corn cultivation like selecting seed and hiring labor—a trend which is more widespread in other regions—the labor in the cornfield is still considered men's work. Both men and women remove the grain from the cob, and in doing so help to select seed to save for replanting. So although men specialize in the cultivation of corn, women in farming households often participate in the selection of seed.

Finally, these stories also illustrate the aging of maize producers. Not only are older migrants and residents largely responsible for corn production, but work experience and salaries in the United States are contributing to the younger generation's preference for work beyond the field.

RESIDENTS' STORIES

Ángela

Ángela had been accustomed to having her own money, because as a teenager she worked as a domestic in Mexico City.[19] Husbands and wives often have different sources of income in San José and may use their income on different things. Men earn money from agriculture and other employment, and sometimes they sell goats. Women, on the other hand, may run a small store out of their home, do piecework embroidery, or own pigs and turkeys which they can sell on occasion for cash. When young children need something their mother often pays for it from her income—and teenagers are generally expected to contribute to the labor and income of the household.

Ángela's husband Raúl is in Oregon working in a restaurant and sends home between $300 and $400 every four to six weeks. Ángela had borne six children, but only four survived infancy. For ten years she had lived with her parents-in-law and as many as a dozen other relatives in a two-room house. A few years ago Ángela and her children were able to move into a new house with three rooms and a dirt floor that had been constructed with Raúl's remittances. Several years after my first interview with Ángela they had purchased a telephone.

When Raúl left for the north the money he sent went to his father, who was in charge of dividing it up and putting some aside for the construction of Raúl's house. In a three-year period he had not yet seen the house his savings had built. Ángela lives in the new home, but she spends part of each day washing dishes, cooking, and helping out at her mother-in-law's house. In late 2002 Raúl visited his family in San José for a month, after which he did not make it back home for five years. He tells me by telephone that he plans on returning to San José to live sometime soon.

Ángela would like to work in a valley maquila, but her absent husband forbids it. Now that her children are in school she thinks that her time could be better spent working for an income. But her husband and in-laws who live in San José insist that she is needed to help cook and watch the children when they are not at school, and that it is not a good idea for a married woman to interact with unfamiliar men, even for work.

Ángela's story points to one trend related to the transnationalization of the household through labor migration, a commodification effect in which unremunerated labor is evaluated according to its monetary worth or its hourly paid equivalent (Lem 1988; Lem 1999, 110). This is not to say that women taking care of their children and cooking for the household believe that they should be paid for doing such tasks. Rather, some women, Ángela among them, compare the time they spend doing domestic chores with the income they could be earning in a factory for the same amount of time. Sanjosepeños who perform unpaid labor—be it housework or subsistence corn production—view their labor as a commodity. Some interviewees argued that their time was better spent in paid employment and attempted to lessen their contribution to the unpaid labor of the household. There was some reluctance to perform unremunerated agricultural tasks, so that households sometimes hired day laborers to farm corn for household consumption.

Ángela's husband, Raúl, is part of a slightly older generation of migrants who send money to finance the cultivation of a small milpa for household consumption (in his case, a little less than a hectare), or to cultivate corn themselves when in town. He first left for the United States in the late 1980s, when he was thirty-five years old. He worked on a ranch in California but later joined his older brother, José, with whom he cleaned hotels and worked as a busboy. The two brothers live together in a house with twenty other people in Oregon and visit San José every few years.

Ángela has a small milpa cultivated once a year, but it is Raúl's father who selects the seed from his own fields, hires the necessary labor, and organizes the irrigation of the milpa for Ángela while her husband is abroad. He hires friends or relatives to help work the milpa, as he does for José, his other son working in the United States. Although two of his sons work abroad, they had considerable experience in the fields before migrating, and both send money home specifically for growing corn. In Ángela's household her husband's remittances and her small earnings as an embroidery pieceworker are spent on household items such as food, healthcare for an ill child, school supplies for the children, and occasional irrigation and labor for the cornfield. Subsistence corn has always been grown by unpaid male labor, but without enough available male labor, households now rely on paid day workers for this work. When her children were a little older as preteens and teenagers, two of them found work in valley maquilas, and one took the trip north to join her father and work.

Juana and Her Family

Not all migrants who work in the United States are able to save money for investments in machinery, cars, agriculture, or construction back home. Juana is in her mid-twenties, one of seven children. She lives with her husband, her children, her teenage brother, and her parents in an adobe house with dirt floors. Juana tells me that her parents are campesinos who grow several types of maize, including elote. Her husband and her brother-in-law have experience working in the United States. Her husband is a construction worker who currently works with his father, before which he was employed for three years in the granjas. Juana explains, "There's no work here, that's why everyone leaves and comes back, leaves and comes back, leaves and comes back" (interview, 26 October 2001).

Juana left school at the age of twelve to stay at home and help out. As a girl she learned how to embroider and worked in her mother's tomato crop on land rented in the town of Altepexi. Shortly after she married, in 2000, her husband went to California with friends for the first time. For a year he was an agricultural day laborer earning at most $1,000 monthly, of which his living expenses came to $450 to $600. After paying off his debts and travel costs he barely had any money to send home. Now in his late twenties, he works as an agricultural day laborer in Chilac.

Juana's sister, Luisa, runs a small grocery store out of her house. Luisa's husband also worked in California, in 1990 at the age of sixteen, but found it too expensive to stay long. He made a mere $2.75 an hour driving a stacking machine in a tomato-packing warehouse, and believes that his employers could pay him whatever they wanted because he did not have immigration papers. For this reason he would prefer legal contract work to undocumented employment. The little money he sent home to San José covered about a quarter of the cost of building the store that his wife now runs. Little by little he has saved the rest by working in valley granjas and cultivating one cycle of corn per year. In 2001, when he was employed at a granja, some of his family members offered to lend him the money to travel to Las Vegas, and to help find him employment once he arrived. As he told me, everyone says that it is increasingly difficult to cross the border because of the terrorist attacks of September 11, 2001, but despite these concerns he left for Las Vegas during the summer of 2003, at the age of twenty-nine. Juana's husband, a few years younger, also left for the United States, but to Oregon.

Neither brother had a good employment experience in the United States during their initial trips, nor did they save much money. Yet both brothers decided to try their luck again across the border. Their case illustrates that although some of the migrants who work in *el norte* are unable to save money for investments back in San José, they still return to the United States because agriculture alone cannot maintain a household and available employment in the Tehuacán area is not sufficiently remunerative. Unlike younger members of the town who dislike maize production and say they will not work in the fields like their fathers, these two men, who are in their late twenties, combine agricultural production with nonagricultural employment. They also worked as agricultural laborers in the United States, a type of work which is now unpopular among younger migrants.

José

Raúl's older brother, José, is thirty-seven years old and a "successful" migrant from San José. Like most Sanjosepeños he had many household tasks and jobs as a child and then as a teenager. Between the ages of ten and twelve he milked cows and sold milk door to door in town. At the age of twelve he started to work in the fields with his father early in the morning, after which he went to school for the day. He left school to work in the fields full-time at the age of fourteen. Two years later he herded goats and at seventeen he assumed full-time employment in a valley granja. On his days off he planted wheat, corn, and tomatoes with his father, for both domestic consumption and sale at markets in Tehuacán and Ajalpan. They practiced sharecropping with water socios.

In the late 1980s José went to the United States for the first time and worked for four months in the tomato fields, where he made $3.25 an hour. He found the job through Sanjosepeño friends. After working as a field hand and gardener in California, he returned to San José, only to leave again for Oregon, where he had heard there were more opportunities. When José arrived in Newport in 1992 he was among the first Mexicans there, a fact he is proud of. By 2001 he estimates that there were more than a hundred Mexicans in the city.

José's first job in Oregon was cleaning shrimp, for which he made $5.25 an hour. He lived inexpensively with friends in a shared apartment and returned to San José once every two years. He then found work as a dishwasher. But in the late 1990s he decided to work on a factory boat in Alaska because it paid considerably more than what he could earn at the restaurant or the fish factory. With overtime José earns about $2,800 a month, and his room and board are included. With this income he has purchased a water share from San Mateo (an auxiliary town) and a tractor that he rents out. He plans to buy a truck for construction next. His wife, also from San José, is encouraging him to be a mayordomo and spend money on the required festivities, but José says he is going to wait until he has obtained the truck. He is a successful migrant in that he has saved enough money to purchase land, a water share, and a tractor, all agricultural investments: "When I had good jobs in the north I sent money to my parents. My mother taught me to save and to think of the future. I would send up to $1,500 a month and my family would put it in the bank for me. My parents and wife would use about $500 a month for food and things. But I also bought four pieces of land, six hectares all together. I have never used Procampo, nor has my father. You've got to be friends with the mayor's office to get it" (interview, 21 November 2001).

José believes that there is a difference between migrants his age (mid- to late thirties), who send money home to work the land or do so themselves when in town, and a younger generation of migrants who neither know how to work the land nor want to return to it. As a successful migrant José exemplifies the benefits of migration in helping families and the town. But he also expressed doubts about the town's future: "What's going to happen to this town if no one wants to work the land?"

Juan

Juan is a returned migrant of twenty-six. He left school when he was fourteen to find work in the valley and help support his seven siblings. In the late 1990s, at the age of eighteen, Juan decided to head north to Oregon, where he had an aunt and a cousin who could help him get settled. He borrowed money from a local lender to pay the $1,600 in coyote fees. The first job he got was washing dishes at a fast-food restaurant. Although he had originally planned to work for three years and return home, he decided to stay longer, in part to learn how to speak English. Juan explains that he spent the first four years in the United States working to help support his parents and siblings back in San José. He sent $200 a month back to his family to help with their upkeep, his sister's university costs, a small milpa, and the cost of building an addition to the house. At his next job, as a waiter, he was able to save for a car and the construction costs of his own house in San José. He also opened a small store that operates out of his house. After traveling back and forth over the years Juan decided to move back to San José permanently in 2004. He now lives in the house he had built with his wife and two children. Since he does not grow corn or work in the fields, he purchases his grain from the market in a neighboring town, where it is slightly cheaper.

Juan's story illustrates the key role of migrant earnings in maintaining the rural household and helping to improve the socioeconomic standing of the younger members of the family. Remittances can free up the labor power of other members of the household: the money that Juan sent home helped with household costs, and his sister was able to study at the university. Juan tells me that migrants help create jobs in town (interview, 8 June 2005): "People in the north help this town by constructing cement houses. The mayor doesn't create any work, but migrants do. They start up businesses, they buy things here. Those that have been in the north buy cement, construction materials, or eat at the local taco stands. They create jobs for this town." His story also illustrates how a younger generation of migrants lacks experience in maize agriculture and fails to take it up once back in San José. Juan tells me that there are several reasons why he does not work in the fields: "First, there is no water here. Second, the corn doesn't grow the way it used to. It's not as productive. And third, there is no place to sell corn anyway."

The conditions for maize agriculture—and by extension, the related in situ maintenance of different local varieties—have become more difficult because of rising costs, declining levels of spring water, and the policies of the neoliberal corn regime. When residents were asked if there was a criollo that they do not plant but that their parents or grandparents planted, I was told that the yellow corn grown largely to feed farm animals, called *co'tzi*, was no longer in use. This is probably because inexpensive yellow corn is available at local stores, or sometimes given out for free as part of government programs (such as PROGRESA, or more recently Oportunidades). Residents also report that the local variety of red corn, *chichiltzi*, used for making *atole* during Holy Week, and the *yahuitzi* blue corn variety are both grown less often than a couple of decades ago. There are few buyers for such corn. Also, until the 1990s many residents grew their milpas in the traditional Mesoamerican fashion, intercropped with beans and squash. This is less common today. The pattern of land use is also changing. As more remittances made their way to San José, the town's inhabited area expanded and the land devoted to cornfields shrank. While elote corn makes up a larger share of overall agricultural production these days, the parcels of land on the edges of town, which served as cornfields just five to ten years ago and were passed down within families to their younger members, are now the site of migrants' cinderblock houses.

As argued by In Defense of Maize, the neoliberal corn regime has exacerbated the difficulties faced by smallholder producers. But rather than interpret maize production as the renewal of a millennial culture, this chapter has demonstrated how the recourse to corn interacts with state policy, economic crisis, and environmental problems such as shortages of spring and galería water.

In the absence of policies which support the food security and livelihoods of rural Mexicans, the neoliberal corn regime defines development as the export of labor through transnational migration or its indirect export through maquilas. Its policies and conditions push young, rural inhabitants into transnational labor circuits; for their part, valley residents devise strategies for surviving or for improving their socioeconomic standing. They have further diversified their income-generating activities, combining corn production, which serves as a social safety net for the older generations, with transnational migration and maquila or granja work for the younger generation. For the most part, migrant and maquila earnings are spent on household maintenance and home construction, and to a lesser extent are channeled directly to agricultural costs or the startup of small businesses.

While this is a local strategy that allows families to adapt and even advance in hard times, the stress and strain on individuals and families is high. Living as undocumented workers in the United States, migrants are vulnerable to deportation and are paid the lowest wages. And while there are migrants who manage to secure better-paid jobs and work permits, the majority are not so lucky. Similarly, maquila and subcontracted home workers make up a highly disposable, low-paid workforce. Even when young workers worry or complain about exploitation and low wages, they tend to return to valley industry or head north—knowing that their access to the work will not last.

Residents turn to corn as a social safety net, but they also doubt whether their future will include agriculture. As migrants return with goods and any earnings that they managed to save, they also bring home their experiences of living abroad and reformulated expectations about how to make a living. These expectations contribute to the monetization of agriculture and to a lack of knowledge about maize agriculture, and a lack of interest in it, among the younger generation.

Chapter 5

FROM CAMPESINOS TO MIGRANT
AND MAQUILA WORKERS?

The paradox of US-Mexico integration is that a barricaded border and
a borderless economy are being constructed simultaneously.
—Peter Andreas, "Sovereigns and Smugglers" (1992)

Sanjosepeños will get across the border no matter how many
guards or police there are.—Maize producer, 22 June 2006

I am a child of God, I am a screwed over child. I am a peasant!
(*Soy hijo de Dios, hijo de la chingada. Soy campesino!*)
—Maize producer, 16 February 2002

You can't make any money in the countryside! There is no money
in the milpa.—Teenage migrants, 30 November 2001

This chapter examines the generational difference in attitudes toward maize agriculture and agriculture-based livelihoods. While
residents of all ages agree that income from off-farm employment,
particularly in the United States, where the wages are higher, is
important for household maintenance and socioeconomic advancement, the older generations describe agriculture as a dignified livelihood. In contrast, migrants in their teens and twenties associate
corn agriculture with their parents' or grandparents' generation and
with tradition, poverty, and burdensome work in the fields. As we
have seen in earlier chapters, maize agriculture, criollo varieties, and
corn-based foods have a complex history of association with class,

race, and gender, as well as ideas about modernity and tradition in Mexico. Southern valley residents navigate dense layers of meaning about maize agriculture. Their own experiences with rural policy and community outsiders, such as nearby city dwellers, government bureaucrats, and the media, contribute to the ways maize agriculture is discussed and experienced. Although the controversy over GM corn is relatively unknown in the town, residents are familiar with the opinions and attitudes of their urban neighbors and of government employees and agricultural extension workers about their agricultural practices and way of life.

Younger residents search for a better life off-farm, much like rural inhabitants the world over who leave the countryside to find work. These southern valley migrants and maquila workers, though, tend to build homes and plan their futures in San José—to establish their families and homes in town.[1] Yet at the same time, they see this rural future as a largely non-agricultural one. As one returned migrant who spent ten years in Oregon tells me: "I think our town is turning into houses. All houses, fewer milpas" (interview, 16 July 2006). It is possible that today's migrants may take up maize agriculture as they grow older—the recourse to corn becomes more important as a male head of household ages and is no longer able to find paid work—but I do not think this likely for most. Not only do young residents discuss maize as an unprofitable crop and show little interest in and knowledge about maize varieties and production, but as outlined in chapter 4, the conditions for agricultural livelihoods are increasingly more difficult.

Whether they are working in the fields, washing dishes in restaurants in the urban United States, or working from home as a subcontracted embroiderer or maquila worker, the daily experience of residents informs how they narrate their lives and their perspectives on the future. The way migrants discuss their lives is shaped by racial discourses and the history of racialized water conflict in the valley, as it is for older generations, as well as through the experience of crossing the border and living in the United States as illegal and deportable workers. Similarly, for female migrants and maquila workers, it is their journey to work as much as their income that challenges local gender norms and influences how they reflect upon their lives.

The Sanjosepeños I met during fieldwork tried to assimilate or downplay their Indianness when in the city or other parts of Mexico.

But doing so is not always successful, and works to reconfirm their experience of being looked down upon as indios. Residents are identified as Indians when they travel based on the way they speak Spanish (or for speaking Nahuatl), their hometown, their dress, the work they perform, or visible signs of poverty. Indians are seen as immobile or more bound to place (the land or their indigenous community) and unable to move through different social spaces unmarked. In dominant post-revolutionary conceptualizations the idea of the mobile Indian or indigenous migrant was in fact a contradiction in terms, since Indians who left their communities and spoke Spanish were no longer seen as Indians but rather assimilated mestizos (Martínez Novo 2006, 75).[2] Similar ideas about the immobility of the indigenous (and other racialized groups) are found in other former colonies (Razack 2002). Adding to regional and local understandings of what it means to be indigenous is the Sanjosepeños' regional reputation as backward, violent Indians mired in the traditions of their community. As explored in chapter 3, residents employ the stereotype of violent Indians in very distinct ways, including as a reference to the social memory of past conflict over water and power or to discuss, proudly, resistance to the usurpation of local resources.

In light of the largely negative associations with being an Indian, it is not surprising that when describing themselves, residents should prefer terms like campesino, migrant, northerner (norteño), worker, Sanjosepeño, and original inhabitant (originario) over Indian (indio) or the more neutral term of indigenous (indígena). I argue below that older residents assert a campesino identity to affirm their humanity, in contrast to both the dehumanizing connotation of the label indio and the neoliberal model for the countryside.[3]

Residents negotiate the dominant and regional understandings about being an Indian differently and in some cases even challenge or reformulate them. For example, a school director and teacher in San José who returned to town after receiving a university education in Mexico City tries to impart pride in the Nahuatl language, symbolism, and regional culture to his students. Although young migrants and maquila workers tend to avoid the label of indio when referring to themselves, much like the older generation, their travel and wage labor experience provides them with an avenue for social prestige, assimilation, and in some cases rethinking the negative connotations of being an Indian—in other words, a means of social mobility. In this sense today's young migrants and maquila workers are more spatially and socially mobile than previous generations, even though Sanjosepeños have a history of traveling within and beyond the valley to sell local bread (*pan de burro*) and agricultural goods, and to work as cane cutters, domestics, and to a lesser extent contracted migrant workers.

Migrants and maquila workers bring home more than just wages; they bring home a different sense of their relationship to the land and of being an Indian, and even new ideas about gender. The experience of crossing the border and working as an illegal is a source of pride in male migrant narratives. And the wage labor experience of female maquila employees and migrants challenges preconceptions about gender roles and proper femininity.

A paradox of the neoliberal corn regime is that the erosion of productive resources for campesinos is regarded in official circles as a form of rural development, but is experienced by the younger generation as freedom from the peasant-based livelihoods of their parents and grandparents—perhaps until the work dries up, wages decline, or they themselves are no longer employable. This younger generation experiences a relaxation of parental and patriarchal authority yet faces exploitation at work as they join transnational, gendered labor circuits.

INEFFICIENT TRADITION VERSUS DIGNIFIED WORK

In the valley, as we saw for the GM corn debates at the national level, there are different versions about the effects of neoliberal policy on small-scale producers. Some officials from the Ministry of Agriculture in the Tehuacán area believe that the potential to modernize valley agriculture has not been realized because of the backward attitudes and practices of small producers. In this they reflect the contempt for the practical knowledge of peasants found in the modernist model for agriculture. The same valley officials spoke well of more successful producers from Ajalpan and other valley towns who experimented with commercially viable crops like broccoli.[4]

In regional government offices some *técnicos*, or extension workers, treat their indigenous campesino clients with derision and condescension. On several occasions I was told that it was hard to alter the practices of campesinos because they were stubborn and unwilling to adapt, or because they believed that the state owed them something. The perspective of these regional técnicos echoes the minister of agriculture's statement, discussed in chapter 2, that the government is "fighting against a culture" in the countryside. Yet the suggestion by técnicos that campesinos believe the state is obligated to them seems in many cases to be accurate. Sanjosepeños suggested to me that the state was failing to live up to its duties, particularly as a source of rural subsidies or support. On the other hand, the técnicos' perspective—that peasants are too stubborn or unwilling to change—ignores an important part of peasants' attitudes. Campesinos may refuse to take up state programs or conform to official expectations, but as both individuals and households they have shown considerable adaptability and flexibility in dealing with crisis. That household members work longer and more intensively, and recruit more family members into the work—"self-exploitation," as it was called during the agrarian debates of earlier decades—is central to rural survival. Sanjosepeños have adapted to environmental and economic crises through recourse to corn and diversifying household strategies. They have changed crops (taking up a versatile variety of corn, elote), incorporated new activities into the work repertoires of the household (including transnational labor migration and maquila or granja work), and interspersed income-generating activities (such as subcontracted piecework) into their other household duties.

Some officials in the Ministry of Agriculture suggest that recent policy alters little in practice even as the situation of rural agricultural producers continues to deteriorate. One técnico, Manuel, with whom I spent time interviewing valley producers, consistently expressed frustration with state programs along with sympathy toward his clients and their plight as peasants. This frustration ultimately led to his renouncing his public sector job.

While the regional state officials whom I encountered discussed Sanjosepeños and other rural producers in ways ranging from sympathetic to paternalistic and disdainful, peasant farmers reject the view that their agriculture is inefficient or should be abandoned. Rather, in everyday conversations they blame environmental problems as well as the government, its policies, and corruption for their difficulties in generating adequate income from agriculture. Older residents often said that they were hard workers in need of water, government help, or employment opportunities. It was not uncommon for residents to ask me if I could help them find work in Canada or the United States. As one *comunero* who stopped me on the street to ask if I could find him work told me, "We are workers here. I may look old, but I am still strong!" (20 May 2001).

On Being Campesino

During a visit to the corner store for a mezcal after working the milpa, one indigenous maize farmer in his late fifties explained the local conflicts over irrigation water by insulting his fellow residents: "Those damned Indians!" he said, "violence is all they know!" He continued on to differentiate himself as a "campesino," but in a manner that hinted at ambiguous and multiple layers of meaning. To be a peasant is a dignified identity based on hard work, but it also implies an ongoing struggle to stay afloat. The farmer's use of "Indian" shows how closely connected the term is to violence, even if violence is a reaction to injustice and disenfranchisement. As he took a swig of his mezcal, the farmer emphatically announced to me and the shopkeeper, "I am a child of God, I am a screwed over child. I am a campesino!" (interview, 16 February 2002).

For Sanjosepeños to refer to themselves as campesinos can signal both a rejection of dehumanizing discourses about Indians and the neoliberal state's abandonment of its previous commitment to the peasantry (See Macip 1997, 2005). A product of the revolution and the cultural politics of the post-revolutionary state (Vaughan 1997; Boyer 2003), the identity of the campesino as the rightful beneficiary of the revolution still resonates with valley residents today. Indeed, as Boyer has argued, the category of campesino as a culturally and politically meaningful identity is "one of the most enduring legacies of . . . the revolution" (2003, 225). It is an ideological construct that signals a way of seeing the world based on shared experiences of political and economic disenfranchisement and a relationship to the land (ibid. 20). An older generation of Sanjosepeños—who remember the struggle for state recognition of their claims to land and water during agrarian reform and later—affirm their campesino identity in several ways: they not only refer to themselves as campesinos but cultivate maize and have turned down offers to sell their ejido land to agribusiness.

In San José the use of the term "campesino" also refers to the ongoing and intensifying plight of the countryside. At a comunero meeting where I was conducting farmer interviews (28 May 2001), a campesino sharing the wood bench with me repeated a phrase I heard many times, "Here, in San José, *we are all screwed!*" A younger producer leaned over and commented, "We used to grow crops that were sold in the city . . . But free trade is screwing us over. Products from the other side [of the border] are already here." When my neighbor on the bench overheard us discussing the scheduled termination of the transitional support payments program (Procampo) in 2008, he turned to me and asked somewhat incredulously, "It's going to end?!" The corn producers I interviewed articulated a range of knowledge about government policy and programs, but most felt that the government was not living up to its promises. Some campesinos, like my comunero neighbor on the bench, were worried about what would happen when Procampo ended and they would no longer have access to the subsidy. Although programs like Procampo are insufficient to absorb the full impact of the economic crisis, these producers felt that government aid had only recently been introduced and made available to residents.

Younger residents hold a variety of opinions about who they are and what their futures hold, ranging from those who portray San José as a town of hard workers, to those who said they would take up agriculture if there were more water or resources, to those who portray agriculture as unprofitable and incompatible with the times. This last view is particularly strong among migrants in their teens and twenties, who speak of their futures in San José but are apprehensive about what that future will look like.

THE PRACTICAL KNOWLEDGE OF PEASANTS

During one of our discussions about local agriculture Don Casimiro, a producer of rain-fed maize, explained that although he gave up farming tomatoes for sale on the regional market, he never stopped growing corn because of how resilient it is as a crop. "The cornfield is a lot less delicate than tomatoes" (interview, 22 April 2001). When I asked what varieties he planted, he told me that "most [Sanjosepeños] use criollo seed for our corn . . . It's been a few years since I bought seed. It was criollo. Well, maybe not a hundred per cent criollo. I don't really know what it was." Don Casimiro had bought valley-grown seed, but his comments signal how seed travels in and out of the valley and that criollos could in fact be the result of creolization between improved varieties and landraces.

In the southern Tehuacán Valley regional markets and stores sell varieties of local, non-local Mexican, and imported yellow corn. Maize producers in San José agree about the main growing characteristics of local criollos, but when discussing varieties available in the market, residents (whether farmers or those responsible for cooking) are more likely to disagree about their growing and cooking qualities. Residents have difficulty identifying a couple of varieties for sale that are used for food or feed (but not seed), whether they are scientifically improved varieties or criollos (such as the "poblano" and "Sinaloa" varieties).

For seed, residents save, exchange, or purchase southern valley criollos from their neighbors and local stores. Most types of corn grown in the valley are criollos—landraces and creolized varieties—adapted to the local climate and soil conditions. Criollos are preferred over hybrids for planting because they perform better.

In San José farmers classify local criollos according to color, maturation rate, size (wide or narrow), and whether they are rain-fed or irrigated. Local and regional varieties have Nahuatl names which refer to their maturation rate. As in much of rural Mexico, where one or two types of criollo corn are favored for local production (Perales, Benz, and Brush 2005), in San José there are two main cultivated varieties. Residents plant "a six-month" criollo of irrigated maize that has a narrow kernel preferred for commercial elote, called *chicuase*. After five months they may harvest the crop for elote. As previously mentioned, the decision to harvest the crop as elote or grain largely depends on the market price for elote and whether there are interested buyers.

The other main corn is a "four-month" rain-fed variety (*nahuitzi*), which can be harvested at three and a half months for locally consumed elote or grown the full four to be consumed as grain. Nahuitzi is not suitable for sale in Mexico City because it has wide kernels and for this reason is generally considered a variety for use as grain. Another variety grown for household consumption in both San José and Chilac is the wide and sweet *macuiltzi*. It can be eaten as elote at four and a half months or dried for grain. Two criollos grown on a much smaller scale, and reportedly on the decline, are *chichiltzi*, a red corn grown for making the cornmeal drink called *atole* during the Holy Week holidays (*Semana Santa*), and *yahuitzi*, the local blue corn, grown for four and a half months and used for making blue tortillas.[5]

A good elote cob should have fourteen kernel rows, whereas grain corn (or "wide" corn), should have eight to ten rows. Residents look for the biggest and healthiest corncobs (*mazorcas*) from the crop to choose seed for the next season of planting. They then shell the kernels from the cobs and store the grain in plastic bags until the next cycle, or for up to a year if pesticide is applied. Saving seed is an individual or household activity, rather than an organized community activity. There is no current or planned collective project for saving criollos, as there is in some other places.

As elsewhere in Mexico, producers from the southern valley try out new seed when it appears in markets or neighbors' fields (Louette 1997). Hybrids are more popular further up in the valley, in towns bordering the city of Tehuacán, where there is more irrigation water along with different soil qualities and larger landholdings. Several Sanjosepeños reported that they had experimented with hybrid corn after seeing larger corncobs growing around the city of Tehuacán and for sale in the city's main market. They had noticed that although the stalks grew shorter, the corncob was bigger.[6] Hybrids also have a reputation for higher yields—residents report hearing this from agricultural extension workers and other valley growers. Producers experiment with so-called modern technologies, and when the technologies offer an advantage (even if it is a social one), they are incorporated into the producers' agricultural practices. Residents do not see any local advantages to growing hybrid corn.

Conceptualizing Agricultural Knowledge

For many years international agricultural research was concerned with the improvement and diffusion of seed varieties and assumed that peasants and small-scale farmers were either like rational scientists whose experiments would lead to the adoption of appropriate technology such as improved seed or ignorant farmers whose incapacity or stubbornness would interfere with the adoption of technology.[7] The research overlooked the dynamism of agricultural practices and their susceptibility to social and cultural influences. But as the study of "indigenous" knowledge (alternatively referred to in the literature as traditional, local, or practical knowledge) gained momentum in the 1980s, agricultural research began to question the appropriateness of technology, and ideas about small farming practices changed (Chambers, Pacey, and Thrupp eds. 1989). Paul Richards (1989) demonstrated that smallholder agriculturalists learn skills to adapt to changing conditions in the field. Each season requires an adaptive "performance" by farmers, depending on factors such as changes in the weather and the availability of household resources for farming. In this way "indigenous" agricultural knowledge is not static but the product of experimentation and performance in various contexts.

Stephen Brush (1996, 4) explains how the concept of indigenous knowledge refers both to knowledge systems unique to minority groups (including indigenous peoples) and to informal knowledge systems which are distinct from "formal and specialized knowledge that defines scientific, professional, and intellectual elites in both Western and non-Western societies." Indigenous knowledge based on oral tradition may be just as systematic as formal knowledge. These definitions of indigenous knowledge resonate with how the concept has been used in international forums and agreements. At the International Symposium on Indigenous Knowledge and Sustainable Development in 1992, "indigenous knowledge" was used synonymously with traditional or local knowledge, to refer to that knowledge which is developed by a given community (indigenous or not), and which is distinct from the international knowledge system based on universities, public research centers, and private industry.[8] As the anthropologists of science insist, however, "all knowledge systems are culturally rooted," including the formal scientific tradition (Hess 1995, 187).

The concept of indigenous knowledge does have its pitfalls. It can be employed to mean knowledge which promotes the sustainable use of natural resources, but this is a generalization about indigenous knowledge (and often indigenous peoples) which may or may not be accurate. In some usages it is wrongly implied or assumed that indigenous knowledge is shared equally or equally accessible to all members of the community. In the southern Tehuacán Valley knowledge about maize is both generational and gendered. Men and women are responsible for different aspects of the crop and its transformation into food. Additionally, the concept of indigenous knowledge can reify the boundaries between such knowledge and the formal, scientific tradition in ways similar to the billiard ball model of culture (alluded to in the Introduction). In contrast, anthropological studies have demonstrated that when local agricultural knowledge—or what James Scott (1998) called "practical knowledge," or métis—is elaborated and handed down from generation to generation, the process is a dynamic one in which both the local and dominant knowledge systems and practices are transformed (Brush 1996; González 2001; Gupta 1998; Nader 1996; Richards 1989). Studies emphasize the fluidity between different knowledge traditions. The practical knowledge of peasants interacts with, and adopts, elements of conventional, science-based agriculture.

In his research on the rapid adoption of a particular type of Bt cotton among small farmers in Warangal, India, Glenn Davis Stone (2007) draws on the cultural evolutionary distinction between "environmental" or individual learning and "social learning" in the adoption of innovations. While environmental learning is based on experimentation, social learning is often based on teaching or the imitation of other farmers. A farmer may imitate other farmers with social standing in adopting certain seeds, rather than taking up the seed because of its traits; in the literature this is called a "prestige bias" (Henrich in Stone 2007, 71). Seed varieties can be markers of modernity or tradition which rank social prestige in the community and beyond (Mosse 2005). Or a farmer may adopt a seed, practice, or technology because others have done so; this is known as the "conformist bias" in social learning. In the southern Tehuacán Valley, while a prestige bias may be at work with the novice planting of hybrids, farmers revert back to criollos because hybrids do not grow well.

Stone argues that individual and social forms of learning are intertwined. Social learning accumulates knowledge based on the results of prior experiments undertaken by other farmers (2007, 84). Although many farmers in India use small-scale experimentation to help evaluate and select seed (Gupta 1998), Stone found that in the adoption of Bt cotton, small-scale farmers did not base their decision on the variety's particular agronomic or ecological features, nor on individual experimentation with the variety. Rather, they were influenced by the practices of larger farmers who had adopted the seed and the seed industry's marketing efforts. In the village of Gudepadd farmers emulated a large-scale farmer who had been recruited by the seed industry to plant a demonstration plot. In other words, in the rapid diffusion of Bt cotton, which was of no particular benefit for small farmers, social learning was not accompanied by or based on environmental learning (Stone 2007, 71). Stone concludes that Bt cotton contributed to an already existing process of agricultural "deskilling" whereby the farmer's ability to perform and innovate was eroded. The inconsistent quality of commercial seed, ineffective government regulation of the seed industry, misinformation about seed, and disruptions in the transmission of accurate information because of the mislabeling of seed contributed to this process of agricultural deskilling.

In the southern Tehuacán Valley the issue is not whether agriculturalists can perform and innovate in the context of a rapidly changing seed market and an insufficiently regulated industry, but whether they can perform in the context of deteriorating environmental and economic conditions and whether the younger generation will learn agricultural skills at all. At this stage of migration and the household life cycle, young migrants have few agricultural skills, prefer nonagricultural work, and view corn agriculture as an unprofitable tradition with few future prospects.

"No hay dinero en la milpa":
Migration and Agricultural Deskilling

When I interviewed male migrants in their teens and twenties (and occasionally in their thirties) they revealed limited knowledge about the details of corn production and the qualities to look for when selecting seed. Two migrant brothers, Donato and Cosme, aged twenty-three and seventeen, are typical young migrants who reported knowing little about maize varieties and agriculture. I interviewed them while they were back in San José after working most of the year in the fish-packing industry and service sector in the United States. The brothers reported wiring money to their parents for their own savings, for their parents' purchase of food and clothing, and for agriculture. Donato had built his own one-level cinderblock house in San José, complete with furniture and a television. Most of the year he lives on the factory boat where he works, and when on shore in the United States he stays at an apartment that Cosme shares with several other Sanjosepeños. For the few months of the year when the brothers are in San José, they eat at their parents' house. Like many others Donato is attempting to establish his own household in San José, a process that may take many years or may never be completed. The younger brother, Cosme, was back in San José indefinitely after a few years up north working for a fast-food restaurant chain. As children both brothers helped with goat herding, worked in maquilas, worked in the fields with their father, and left school around the age of fourteen to work full-time. Neither can describe the details of corn cultivation: the timing of irrigations in the crop cycle, the criollos best suited for different soils, and so on. Like other migrants their age they are fluent in Nahuatl, but cannot remember the names of the main varieties of maize. They told me that agriculture has no future since "you can't make any money in the countryside! There is no money in the milpa" (interview, 30 November 2001). Indeed the majority of the young migrants with whom I spoke claimed that they would never return to the fields—the work is difficult and the financial return small and irregular.

One migrant who left in the mid-1990s at the age of eighteen told me that he had little experience in the fields with his father before leaving for the United States. In response to my questions he said that he did not know the names or characteristics of local maize varieties and soils, or the timing of irrigation: "I don't have a clue how to grow corn or about the varieties. I have no idea about the soil. It all looks the same to me. . . . I'm not sure whether to stay or to go back to the United States. I'm kind of scared. I'd have to go back to make more money. I was thinking of going back to school here, but I don't know. I'm not sure. I've been in San José for four months now, doing nothing. Maybe I'll stay permanently. I have savings for expenses and the house [I'm building], but no job. I'm worried about running out of money" (interview in English, 16 July 2006). As he continued to describe his worries about the future, he explained that he did not consider agricultural day work an option because it did not bring in enough money for the amount of effort expended.

An older male migrant, aged thirty-seven, who had worked in factories and the service industry in California, Las Vegas, and Oregon, similarly could not identify any of the maize varieties grown locally: "I don't know the names of any criollos, I don't know when to irrigate the fields, what type of soil we have here, or even how to save seed. I haven't worked in the fields in at least fifteen years. I don't think it's a good thing that so many young people go north, but since there is no water. . . they need to go." When asked whether he would like to grow crops if he had water, he replied: "If I had water, I would probably grow crops, but it is very expensive. The future here will be difficult; you need a lot of water to grow crops" (interview, 16 July 2006). This older returned migrant differed from younger interviewees in that he expressed an interest in taking up agriculture if he had the resources.

When migrants would stop me from finishing my list of questions about local varieties and soil qualities because they did not know the answers, I would ask them why they thought their older relatives or townspeople grew corn. I was told that corn was grown as a custom or tradition and to produce food, but that corn was definitely not a means to make money: "People grow corn here to eat, so they don't lose the custom, or to get out of the house. They don't grow corn to make money. It's not a business" (interview with twenty-nine-year-old male migrant in English, 8 June 2005).

Since young men can always learn how to farm corn later in life, it is perhaps more significant that many young interviewees considered an agricultural life arduous, unreliable, and traditional. They prefer nonagricultural employment in the United States because it offers less physically arduous work, a quicker return than waiting to sell a harvested crop, and once a job is found, what they see as a steadier income (at least for the duration of their employment). In contrast, male migrants now in their thirties and forties were more diverse in terms of the type of work undertaken in the last five years. Some were both migrants and corn farmers, like José and Raúl in chapter 4. Others had recently worked in construction, truck driving, and other agricultural paid work or sharecropping. Most men of this age group had at some point in their adult life worked in the milpa and in the United States, whereas younger migrants have less experience in the fields. As boys these young migrants may have helped their fathers to farm, but by their early teens they were working full-time in valley industry or starting the journey north.

The strategy that young people pursue of contributing to their parents' household and establishing their own home in town takes them away from maize production and the related agricultural knowledge. Young migrants prefer the short interval between paychecks to the delayed and unreliable profit of the harvest. This is an age group of men who in previous generations would have been skilled in maize agriculture, but who now learn different skills as migrant workers in the service and food-processing sectors in the United States. Unlike in other regions of Mexico, where women become responsible for maize production and take up work in the milpa, in San José the milpa is still very much a male domain.

RESIDENTS' VIEWS OF MIGRATION AND MAQUILA WORK

While older men and women reaffirm their campesino identities, residents of all ages tend to view the inflow of goods and money from off-farm labor as a necessity or a sign of progress and economic advancement for their town. A couple of interviewees also suggested that migration is a beneficial process because it brings new, progressive ideas to town, particularly ideas about the importance of education and about women as equal partners in marriage. Residents often associated the economic benefits of migration and off-farm employment with what they perceived to be social problems: the separation of families, the danger of undocumented border crossing, male migrants abandoning their families financially and emotionally, the decline in available and knowledgeable agricultural labor, teens' disrespect for their elders, teens' use of drugs and alcohol, and promiscuity or the appearance of indiscretion among young female maquila workers. A few of these ideas about social problems will be looked at more closely.

In discussions about migration, young residents compare life in San José to life in urban Mexico and the United States by employing a narrative of "freedom" or of being "free." Guadalupe, a mother in her twenties who was previously employed in Mexico City and is now living in her parents' household while working at odd jobs in San José, told me: "We're so much freer here than in Tehuacán or somewhere else. If we want eggs we walk outside and get them. Look, if we want chicozapote [a fruit] we go pick them. In Mexico City you have to buy everything. Everything depends on money. And you're so restricted, cramped in. I never feel closed in here, but in Mexico City I do" (interview, 28 February 2002). Although the sentiment that residents are "free" in San José is commonly expressed by young migrants, Guadalupe's perspective is unusual. First, she focuses on the ability of the rural household to use goods from its own milpa or yard rather than depend entirely on money and the market for food or an income. This is a perspective (and practice) more common among older residents who value food self-sufficiency—or the ability to produce their own food either at the household and inter-household level through sharecropping—and access to the land, water, and labor resources necessary to make self-sufficiency possible. Despite Guadalupe's emphasis on the quality of life in *el campo*, four years later she told me that she wanted to move to Tehuacán to find work once her husband returned from the United States (interview, 23 June 2006).

Second, the narrative of "freedom" used by migrants is often a gendered one. The way Guadalupe discussed her freedom in San José differs from the perspective of three migrant women with whom I spoke who felt "freer" in the United States. Young male migrants, in contrast, told me that although certain services, wages, and employment opportunities were better in the United States, they had more freedom in San José; and that in town (and in Mexico more generally) they were free from the worries of working without papers and crossing the border, the higher costs of living in the United States, and even stricter laws that regulate activities like the playing of loud music. As one migrant in his early thirties put it, "I worry less here—there is no rent or bills to pay. There is no *migra* [immigration or border police] to worry about. It is freer here" (8 June 2005). These workers are free from the everyday worries of deportability when back in San José. Further, for migrants in their teens and early twenties San José is no longer simply their home-town, but the place they go on holiday to relax and see family and friends. For this group the countryside is a place of leisure.

Other residents viewed the freedom enjoyed in San José as having both positive and negative characteristics. A female storeowner questioned the value of "freedom" for Sanjosepeños in a situation where there is not enough work or food: "We are free [here], but there is almost no food or work" (interview, 31 May 2001). A migrant in his early thirties, who had recently returned from five years of working in the United States, said that life was good in San José as long as one did not think about what the future held: "Life here [in San José] is good for living. That is, only if you don't analyze things, if you don't think about your children or the next generation. We have food but what are we going to give the next generation? There are more possibilities in the United States. I learned to be a mechanic, I learned a little English. There are public libraries! We don't have that here. People should know a bit of the world. Because of this, people here are closed minded" (interview, 2 February

2002). The migrant was also critical of what he saw as "closed-mindedness" in town and residents who did not know about the world through training, education, or travel.

Adult residents also complain about wage-earning teenagers in San José as having a new level of social and economic independence. As contributors to their household income, teenagers may feel that they can ignore parental authority. One of the migrants in his thirties quoted above told me that he thought teenagers who worked in valley maquilas and in the United States did so not only out of economic necessity but because they wanted to have access to consumer goods, to avoid school, and to liberate themselves from parental authority: "I think that more people will go north because of economic need and because many don't want to study. [Going to the United States] is not always for necessity. If they study, it would take them eight years. But young people want to buy clothes, to go to dances. And, if they finish their studies, where are they going to work here? I would say, only 10 percent of young people study. They start at the maquilas but they are introduced to problems there. At fifteen or sixteen they give half their earnings to their parents. And with the other half, they go out. Their parents don't say anything because they like having the money. If parents say something then their kids will stop giving them money. So the kids feel free" (interview, 8 June 2005). Similarly, a migrant who returned to San José to live after a decade in the United States said that parents did not intervene as they should in the lives of their wage-earning teenage children. He believes that migration further contributes to the problem because it results in absent parents, particularly fathers: "Migration helps out financially but it tears families apart. Husbands leave their wives and the kids don't know their fathers. They don't have that family value; everyone needs a dad. Kids who start drinking young don't prepare themselves [for the future], they think they can go to the United States and not study. There are thirteen- or fifteen-year-olds who go out and drink. I wouldn't let my kids under fifteen roam the streets at night. They don't have parents, or a father. They don't go to school, but they don't care because they will just go north to the United States for work" (interview, 16 July 2006).

While some young people may indeed undertake migration in order to avoid studying or to be able to purchase fashionable clothing and other goods, and may have some newfound influence in family decisions, I came across migrants who were pressured by their families to leave school and find employment either in the valley or across the border. Adults have a tendency to view teen behavior as problematic, or different from that of the previous generation, so they may overlook how teens face pressure from their adult relatives to migrate or work. Some of the same peasant families who worry about the agricultural future of San José, and the decline in available and knowledgeable agricultural laborers, encourage or oblige their teen children to work off-farm.

CONSUMPTION AND COMMUNITY

Remittances are largely spent on individual and household priorities. Although a group of Sanjosepeño migrants made a collective financial contribution to repairing the town's church, there have been few instances of migrants pooling resources for a town project or for the benefit of the town as a whole. In this sense migrants often focus on the possibility of economic advancement and social mobility for themselves and their households and do not necessarily work together to transform the collective experience of poverty (Basch, Blanc Szanton, and Glick Schiller 1995; Rivermar Pérez 2000, 92). At the same time, the shared experiences of working abroad or in valley maquilas are changing the social and physical landscape of community in San José.

Migrant income help raises the local standards of consumption, notably with the construction of cement block houses instead of adobe ones, in some cases on former agricultural land. Most homes have electricity (first introduced in the 1960s) and a television but no running water or a telephone—although cell phones have spread quickly. It is increasingly popular to cover dirt floors with cement, but not everyone can afford to do so. There are now two-story homes in town—a sight that residents point out with interest.

Migrants return to San José loaded down with goods brought from the United States and with savings to spend. Some of the goods most commonly purchased abroad or bought with migrant savings in Tehuacán are music, clothes, videos, electronics, and vehicles. The noticeable infusion of goods and cash in town has had an effect on *compadrazgo* (ritual kinship) and fiestas.

The practice of ritual kinship is undergoing modifications as young men and women have access to disposable income. In the past ritual kinship established through and around weddings entailed having a *padrino* (godfather) for the music and for alcohol. But recently migrants have returned to San José with new expectations about wedding parties, expectations based on what they have seen in urban Mexico, in the United States, and on television or in other media. With new consumer expectations about what makes a successful wedding, more people are enrolled as ritual kin to furnish the various components of the party. There is now a *padrino* of the store-bought wedding cake, a *madrina* of the store-bought wedding decorations, a *padrino* of the wedding video, and so on. The ritual kin are seated at a place of honor in the wedding party for their participation in the wedding.

In addition, a generation or so ago families did not celebrate *quinceañeras*, the "sweet fifteen" parties for young women popular in much of Mexico, while they are common today. One female store owner described the change: "Quinceañeras started here in the 1990s. We didn't have them when I was a girl. Now young women work in the maquilas and with that money they save up to have a quinceañera. They also have padrinos, like padrinos for the quinceañera dress, the mariachis, the decorations and so on. Weddings also have more padrinos these days. There are padrinos for the ring, for the music. When I was young, we had padrinos [for weddings] too, but not for some many things. They contributed very little" (interview, 18 July 2006). Most interviewees discussed the ability of maquila workers and migrants to contribute to these celebrations and the higher level of spending on goods and services as a sign of progress—although I did meet a few residents who were critical of lavish spending on fiestas.

At the same time that ritual kinship has expanded to meet the increased consumption of consumer goods associated with throwing a wedding party and the newly celebrated quinceañeras, it appears also to have thinned. Even as the relationships of compadrazgo have proliferated in the planning of weddings, quinceañeras, and first communion celebrations,[9] the dense social networks that characterized sharecropping and ritual kinship and bonded families and households of different resources and prestige have been on the decline since the 1990s. As sharecropping has decreased since the 1980s and agriculture has become more strictly a business, many of the relationships appear to carry fewer of their former social responsibilities and economic obligations. For this reason it appears that ritual kinship has been both refunctionalized and thinned—a topic for further research.

What is clear, though, is that as in other regions, migrants (and maquila workers discussed below) enjoy new social status in their hometowns. The money that migrants earn abroad enables them to build and improve homes in Mexico, buy consumer goods, and participate in local celebrations (Goldring 1999; Rivermar Pérez 2000). Jones (1992, 507) suggests that migrant income and purchasing power translate into a new "migrant elite" whose prestige comes from "wage labor earnings rather than from land, commerce, social status, and political pull." To an extent this is also evident in San José, where younger returned migrants gain social prestige from their purchasing power and experience abroad rather than from their water shares, participation in ritual kin for agriculture, or affiliation with political factions, as was true of older Sanjosepeños and even older migrants who are also farmers.

The new avenues for social prestige exist alongside or help to refunctionalize traditional ones. Returned male residents reintegrate themselves into the community, for instance, by making financial contributions to the faena (as household heads) and participating in traditional celebrations like the mayordomía—although such traditions are also questioned by residents. I encountered two returned migrants who had become mayordomos, a responsibility which involves considerable expenditure. One year of service to the mayordomía costs between $2,600 and $4,400.[10] While such forms of reintegration are expected for migrants with wives back in San José, they do not appear to be expected of teen migrants.

Border Crossings and Masculine Status

Migration from San José increased at a time when undocumented border crossings into the United States were becoming a more dangerous endeavor. In an attempt to circumvent new enforcement efforts at the United States border in the 1990s, Mexicans trying to enter the United States without work or immigration documents redirected their crossings to more isolated and dangerous areas. As a consequence, the number of deaths due to dehydration and sunstroke among illegal crossers rose. Despite increased spending by the United States on border policing and barriers, Mexicans were making more attempts to cross the border illegally.[11] For some young men in San José the risks and dangers of crossing the border, and of finding work in a foreign country, are a source of pride and even local status.

Residents discuss migration both in relation to the dangers of the trip and the potential rewards, which for many residents justify the risks. One morning in her partially open-air kitchen, as her husband Don Ernesto arrived for lunch and a nap after working the milpa, Doña Graciela told me, "One of our sons who is in Las Vegas working has had a lot of trouble crossing the border." On one of his first trips across the border, "They caught him, beat him, and took all his money." When I asked whether coyotes or someone else was responsible for the robbery and the beating, Doña Graciela was not sure. She went on to say that one of her grandsons had also had problems getting into the United States. He was forced to hide in the scrublands underneath bushes and trees to escape detection from the border police. "He didn't eat for four days and barely spoke when he came back." But despite all of this, Doña Graciela said, "It is worth crossing the border because of the money you make—if you can take advantage of the situation and not waste it on drinking. Maybe you could help my daughter across?" (interview, 7 May 2001).

During my fieldwork there were only two accounts of local migrants who had died abroad, and these were because of alcohol consumption and brawls rather than the border crossing. Residents told me that they or their relatives succeeded in their attempts to navigate the border and persisted when they were not successful, because they were determined and used to hard work. One resident even suggested to me that residents' participation in religious pilgrimages to Oaxaca helped train migrants for the difficult journey across the border. One maize farmer explained how residents were determined to find entrance into the United States. He saw their experience working in the heat of the maquilas or in fields as preparation for the border: "Sanjosepeños will get across the border no matter how many guards or police there are. We will do it. We can handle it. We're used to hard work. If it's a kid who tries just out of school, he won't make it. But if he has the experience working the fields or in a maquila for a couple years—where it may not be sunny but it's hot—he'll make it" (interview, 22 June 2006).

The comments and stories about female migrants that I heard from residents and migrants themselves also discussed the resilience required for the journey, but focused more on women's potential vulnerability and victimization at the border.

A few of the male migrants with whom I spoke bragged about the dangers of the border and their ability to live in the United States as undocumented workers. During a discussion with Domingo, a former migrant in his late thirties, he proudly revealed—unrelated to any question I had asked—that he had spent time in an American jail. He had worked in the United States in the late 1980s and again in the 1990s. Like other migrants of his generation, he also grew corn. As we sat on the porch of the ayuntamiento where he now worked, he produced from his wallet a United States correctional identification card. Domingo explained that he had spent almost a month in jail in Oregon for a driving violation. I was impressed with how proud Domingo was of defying the law in the United States, particularly since at the time of our conversation he was employed as a local police officer (interview, 4 June 2001). He kept a record of the jail time in his wallet, like a memento that conferred bragging rights. I also met younger returned migrants who told me they were "wetbacks" (mojados), a term which highlights their undocumented status in the United States and has been used historically in a derogatory manner. Others I spoke to preferred the label northerner (norteño).

For some male residents, labor migration to the United States is not simply a question of fulfilling their manly duty to provide for their household; the risks of the journey and of working without permission in the United States offer new tests for male bravery and resilience. For some residents, notions of male bravery may even be bound up with racialized discourses and their regional manifestations about Indian ferocity (as discussed in chapter 3)—a reappreciation and redeployment of those stereotypes about Sanjosepeños in a way that generates masculine pride and status.

Both young men and women express pride in earning an income, but at the same time they are fully aware of the exploitation involved and complain about the conditions under which they work.

GENDERED SPACE AND THE
JOURNEY TO WORK FOR WOMEN

Movement in, and narratives about, different spaces are an important way that gender is experienced and constituted in the valley. The ability—or inability—to travel and move in and through different social spaces is not only related to residents' racialized ethnicity but to practices of gender as well. Sanjosepeños discuss appropriate femininity in terms of restricted or monitored movement. A woman who is considered a good wife or mother, respectful of tradition, is not to travel unaccompanied. Of course women do travel for work or to the regional capital to take their children to the hospital or withdraw earnings and remittances from the bank. Moreover, for many years women have been agricultural vendors at valley markets and day laborers in the garlic and tomato fields of neighboring towns. They made (and continue to make) pilgrimages to religious sites in Oaxaca, have been employed as domestic servants in Mexico City since the 1950s, and increasingly work in nearby maquilas during the day or even migrate with their husbands or relatives to the United States. However, women of a particular age and marital status are expected to be accompanied by male relatives when they travel (or, if not by male relatives, by children). Today young women's work in maquilas and migration to the United States challenge preconceptions that women's honor and sexuality are compromised through unaccompanied travel and the implied interaction with unknown men. At the same time, certain traditional views of women's comportment and work are reinforced.

Doña Lucía

In the front room of her home, which had been converted into a store to sell snacks, Doña Lucía, a woman of seventy, told me about what marriage was like when she was young. She was married at sixteen in 1948. Her father arranged the marriage and she met her husband for the first time at the entrance to the church on their wedding day. She was dressed in the white wedding gown that was in style in those days, but wore no shoes. Her husband, a local man, was employed as a truck driver who transported crops to Mexico City. He traveled frequently, she tells me, but "women didn't leave their houses then." Since Doña Lucía was for many years a vendor of fruits and vegetables at the regional markets, I asked if she stopped selling in markets after she married. She explained that before marriage she went to the market with her father, and after marriage she continued to go, but accompanied by other relatives. Although her husband "drank himself to death" when she was still a young women, Doña Lucía explained that she never remarried. In San José, she told me, "you are married until God decides." When I asked Doña Lucía and her younger neighbor Leticia what happens if a man does not live up to his responsibilities as a husband or father—for instance, if he is a migrant who stops sending money home—Leticia responded, "There is no divorce here. It's whoever dies first. Here women put up with everything. He can hit you, can have other lovers, can stop sending you money, but when he returns home, you are still his" (6 June 2001).

Women who had been *muchachas*, or domestic servants, in Mexico City during the 1950s and later found their jobs through relatives and were under the care of the employer. Since domestics lived in the house where they worked, their honor was entrusted to this new household—although of course they were vulnerable to the sexual and physical abuse of their employers. Despite being under the tutelage of an employer's household, women who worked as domestics were the subject of rumors in town about compromised honor. Catalina, a woman in her thirties, told me that a few of her in-laws tried to convince her husband that he should not marry her because she had worked as a domestic and therefore was considered promiscuous. It is common knowledge that domestics had Sundays off and would meet each other in Mexico City's big urban parks to catch up, share news of home, and possibly meet a young man to date. Young men from San José employed in the city would also visit the park on Sundays—in fact this is how Catalina met her husband. The patriarchal double standard considered these park meetings suspect behavior for young women but not for men.

Today the rules of courtship are nowhere near as rigid as they once were, and young Sanjosepeños freely choose their spouses. But when women marry, they move into the home of their in-laws and are under the tutelage of the mother-in-law until the couple can afford their own home. Men are still considered the family patriarch, household head, and main breadwinner. Women's income is seen as supplemental.

Women's Roles Reconfigured and Reproduced

The journey to work for female migrants[12] and maquila workers provides this generation of women with the sense that they are more independent of the authority of male relatives than their mothers or grandmothers were at their age. Their access to a paycheck enables them to make a regular financial contribution to the household—which is particularly important when they have children or delinquent husbands—and to have pocket money for spending on clothing, bus fare, meals outside the home, and other small items. Teens save up for their quinceañeras and to contribute to those of their friends. Since this is a celebration practiced among mestizo Mexicans and new to San José, it is clearly a source of status for young indigenous women.

Women's income is also perceived by residents—in both positive and negative ways—as a source of leverage in making household decisions and a source of female independence from their husbands or fathers. As noted elsewhere in Mexico, wage labor outside the home improves "women's ability to negotiate 'patriarchal bargains'" (Goldring 2001, 507).[13] When I spoke to three returned female migrants and a migrant husband about what they thought of their time in the United States, they all commented on the positive attitudes toward women abroad, including an acceptance of women who travel alone, drive cars, and work. They all contrasted these positive attitudes toward women in the United States to what they saw as double standards and more restrictive practices for women in the valley. Women see the status that they enjoy while working and living in the United States as better than their status in their hometown. But I also encountered residents, usually male, who felt that the leverage in household affairs that women gained through wages was a negative development, as illustrated by Ángela's husband, discussed below.

Gender roles are challenged and reconfigured in the southern valley as young women travel, earn a paycheck, and question traditional norms (such as courtship), but at the same time aspects of these gendered roles and expectations are reproduced. Anxieties remain about women who interact with strangers. In the words of one grandmother, young maquila workers have to be careful "because sometimes they meet boyfriends and get pregnant" (17 June 2001).

Maquila work—whether in the factory, in the workshop, or at home—is one of few options for women to generate an income. In contrast, for men maquila work is considered a temporary option, particularly since their main goal is to head north (Flores Morales 2008). As one young man said to me, "How can you save money for the coyote when you're working in the maquilas?" (5 December

2001). Similarly, a returned migrant interviewee with several years of work experience in the United States told me that he would not consider maquila work. When I asked whether he would consider getting a job in a valley maquiladora he replied: "There's no way I'd go to work in a maquila. I'd be a slave there. They don't pay enough money. I wouldn't work as a mozo in the fields either" (interview in English, 16 July 2006). Although teen girls are now expected to work in maquilas until they marry or have children, and after marriage these women undertake a variety of income-generating tasks, the common idea is that married women or mothers should not work outside the home. To illustrate how residents discuss the journey to work for women, I now return to two of the household cases from chapter 4.

Juana and Her Family

Although Juana has never worked in a maquila, she left school at twelve years of age to stay at home and help out. She says that because wages are low in the valley, "It's worth it for girls to work in the maquilas, but not for guys. Guys can make more money elsewhere" (26 October 2001). As Juana's explanation illustrates, in San José the gendered ideology of the male breadwinner is not only found among residents but is clearly at work in the gendered division of labor and the pay differentials of the maquila industry.

As a teenager, Juana's sister Luisa worked as a domestic servant in Mexico City for three years. Now in her twenties, she owns and runs a typical corner store that sells processed foods, grain, animal feed, pop, beer, and locally made alcohol. At seventeen Luisa also worked in a maquila for eight months. She did not like the work. "There was always a huge pile of pants to finish every day," she says, referring to the pressures of the daily quota system. Like other young residents Luisa never got officially married in church because of the expense. She tells me that young people "get together" (se juntan) when they are teenagers, young women at around fifteen while young men might be a few years older. The rules of courtship are much more relaxed than they were for her mother's generation or in the late 1940s, when Doña Lucia met her husband for the first time on her wedding day.

There are rumors about Luisa taking lovers while her husband is in the United States, but the lack of a church marriage is not the source of these rumors. Rather, it is Luisa's practice of traveling alone to other valley markets and to the city to buy stock for her store. While many residents view young women who travel to work in factories or to the United States with their husbands as a positive development, anxieties remain about women who interact with strangers beyond the watchful eye of their relatives.

Ángela

Ángela would like to work in a valley maquila or live with her husband Raúl in the United States. As a teenager Ángela had a job as a domestic in Mexico City. She only went home to San José once or twice a year for a visit during the holidays. Sometimes she and her sister would put their money together to send to their mother. Today Ángela works irregularly as a pieceworker who is given prepatterned cloth to embroider in her home. She often asks her husband to take her and the kids north or to send money so that they can cross the border. But he says this is too risky and does not want Ángela to get a job: "He tells me, 'I don't want to bring you here. The women are free here in the north. They can maintain themselves. You wouldn't need me. You depend on me now because I send you money.' I want to work in the maquilas but he won't let me. He says those workers all have lovers" (interview, 5 February 2002). Ángela's husband also did not allow his teenage daughter to work in the maquila until she turned fifteen. Then he acquiesced.

The allocation of labor in Ángela's household is shaped by attitudes in town about the responsibility of mothers to be at home for their children and about the compromised sexuality of women who work in factories. Although Ángela compares her unpaid labor to potential wages—the commodification effect discussed in chapter 4—in the end her unpaid labor is partly mobilized through a patriarchal family structure. She would be considered a bad mother and wife if she worked in a maquila, because her children are still young and her husband is particularly strict. Ángela's story is interesting, because although she abides by her family's wishes she would prefer to work for wages in a maquila. For the most part, when I spoke to mothers about motherhood and maquila work, they felt strongly that for young women in their teens and early twenties it was beneficial to have this type of employment, but not after marrying or having children—unless the work was subcontracted piecework that enabled them to stay home. Subcontracted maquila work or piecework embroidery is sometimes preferred by workers' husbands (like Ángela's) because it does not require women to leave their home or their domestic responsibilities.

While women see the work as an economic necessity and residents often say they are glad to have it, maquila work is repetitive and low-paid. This again is paradoxical: women are subjected to labor exploitation in the maquila industry, but at the same time this experience and income provide them with a sense of independence, with freedom from some of the social constraints that their mothers faced. The journey to work for women and the export of female labor (as workers in maquilas and less commonly as migrants) challenge normative ideas about gender. Yet this journey to work depends on the maintenance by the maquila industry of such ideas about gender roles, and it also generates anxious discussion among residents about appropriate behavior for women.

"PEOPLE SAY WE'RE INDIANS, [EVEN THOUGH] WE ALL HAVE THE SAME BLOOD"

There are residents—migrants and nonmigrants alike—who question negative stereotypes about Indians based on their own experiences, reflections, and exposure to positive ideas from various sources, such as the local teacher (mentioned above) and media coverage of the growing indigenous rights movement in Mexico, including the EZLN uprising in Chiapas and its Zapatista Caravan of 2001, which stopped in Tehuacán. For migrants, however, their experience abroad also informs their sense of being indigenous Mexicans.

In the United States, Mexican workers are inserted into the racial hierarchy as inexpensive, disposable, and deportable labor (De Genova 2005), yet migrants from San José sometimes experience working in *el norte* as an improvement in their social status and self-perception—in the United States they are seen as Mexicans and not Indians—despite the exploitation and racism. Additionally, this experience, along with wages, work experience, clothing, and the ability to speak English, are steps toward assimilation—not necessarily complete or successful—when back in Mexico. According to accounts by returned migrants, employers, customers, and residents in the United States tend to view Sanjosepeños as homogeneously Mexican. A group of teenage migrants told me that in the restaurant kitchen where they worked in Las Vegas, even though they often spoke Nahuatl to each other, their employers referred to them as Mexicans. Yet when they are in Tehuacán—historically known as the City of Indians—they avoid speaking Nahuatl because they do not want people to think that they are indios. It is just "too embarrassing," I was told (interview, 12 November 2001). One young man from the group explained that in Tehuacán, "people say we're Indians, [even though] we all have the same blood." Ironically, these young migrants may feel more a part of mainstream Mexican society because of their experience in the United States than those who stay behind.

Studies of migrants from other indigenous regions of Puebla have also found that their experience in the United States provides a path toward assimilation or "de-Indianization" when back home (Rivermar Pérez 2000)—but again, assimilation is often not successful or complete. As I suggested in chapter 3, the ways residents engage notions of being an Indian need to be understood in a historical, political, and economic context rather than seen simply as internalized racism. In Carmen Martínez Novo's research on the border between the United States and Mexico, she found that internal migrants who were Mixtec preferred to avoid stigma and discrimination by blending in (2006, 116). She argues that we need to pay careful attention to the "disadvantageous consequences for workers with stronger Indian identification" in Mexico (2006, 54).

The experience of migrating to the United States builds the confidence of some Sanjosepeños. But for those who do not learn English, their experiences abroad may confirm their insecurities about traveling and interacting with people outside the valley. Several migrants reported that if you do not speak English in the United States, you will be ignored, get lost, take the wrong bus, not find a good job, and so on. Not all male migrants preferred working in the United States over staying in San José or were successful in saving part of their income. But as discussed in chapter 4, without other remunerative employment options, many are compelled to return to the United States regardless of whether they had a positive time there.

Migration is an experience that enables some valley residents to think about their racialized ethnicity in a more positive light. In the same group of teenage migrants who spoke Nahuatl to each other at their job in Las Vegas, one young man had a tattoo on his arm: a muscular, heroic cartoon image of a head-dressed American Indian. This young man's identity as an Indian was refracted through his experience of working in the United States as well as through the lens of romanticized media images of indigenous peoples. On the windows of a couple of local vans that take people from San José to the neighboring town of San Sebastián (where residents can transfer to a valley bus that goes to the city of Tehuacán), there are cartoon images of sexy American and Mexican Indian princesses, such as Pocahontas and Iztaccíhuatl.[14] Some migrants have a growing sense of pride in being Indians, understood in part through their experience of being identified or marked not as Indians but as Mexicans, and they may also view being an Indian through the lens of popular images in the media. One's sense of self depends on navigating and translating experience from these different contexts and moments in the life cycle.

While some young migrants return home to San José and re-integrate into the community—as young husbands and fathers, faena participants, and so on—other migrants wear their experience abroad as a badge of honor, differentiating themselves from their nonmigrant relatives and peers. Sanjosepeños complain about migrants who return home acting high and mighty and speak English among themselves as a way of excluding others. These migrants reinforce the differences between themselves and nonmigrant residents, and this differentiation between norteños and campesinos is sometimes understood and expressed in racialized terms.

Young migrants and maquila workers use part of their salaries to purchase fashionable maquila-made clothes, as one way of differentiating themselves from "traditional" Indian residents. Just as a younger generation of migrants avoids the agricultural life of their parents, so too do some young residents shun the wardrobe and comportment of their elders. Most men in San José wear trousers, woven hats or baseball caps, sandals, and button-down shirts that are often stained from work in the fields, while a few elders may still be seen barefoot, dressed in traditional all-white cotton pants and shirts. Young men who want to differentiate themselves from the older generation or who simply think of themselves as just following the latest trend in urban Mexico or the United States wear baseball caps and baggy jeans. They may wear bandanas on their heads in the "*pocho*" or "*cholo*" style and sport tattoos. While this sense of style distinguishes these young men from other Sanjosepeños, it does not necessarily make them mestizos in the eyes of Tehuacán or Pueblan urbanites.

Women are generally dressed in sandals and a dress (or skirt and blouse), covered by a store-bought apron (*mandil*) or shawl (*reboso*), with gold loop earrings and braided hair. According to several female Sanjosepeños, if a woman does not wear a shawl or apron to cover her chest she will be criticized as vain or as sexually promiscuous. Similarly, women in their thirties told me that it was improper and vain to wear makeup in San José, although young maquila workers generally do. Each town has a distinctively embroidered traditional blouse, a particular style of *huipil*, but only a few elderly women still wear them. Teenage girls who attend school or work in maquilas sport stylish, store-bought shoes, dresses, and pants and leave their hair unbraided. Some use makeup. Others combine elements of the traditional dress with fashions from Mexico and the United States. One migrant woman who returned to San José from the United States to raise her daughter wore braids, gold earrings, and a mandil but with shoes instead of sandals. Many "older" migrants in their thirties worked the fields but wore baseball caps instead of a woven hat, or jeans instead of slacks.

Regardless of whether members of the younger generation are intentionally making a statement by what they wear, their fashionable attire and consumer goods, along with their off-farm employment, mark a difference between generations and between "tradition" and "progress." Residents juxtapose "traditional" or "closed people" (*gente cerrada*) with the "open-minded" and "forward thinking." Migrant experiences and consumption patterns, and the inflow of U.S. dollars, are discussed in terms of both progress and its social costs. For young migrants, their parents' or grandparents' way of life, including small-scale maize agriculture, is seen as fixed in the past and out of step with the times.

Young residents who work in service jobs in the United States or valley maquilas distinguish themselves from their parents and grandparents, who worked in agriculture and led lives steeped in hardship and irrigation conflict. For migrants, their income, travel experience, English-language ability, and greater access to nontraditional clothing and other consumer goods enable them to cast off the negative associations of being indigenous or to redeploy them.

As Friedlander found in her study on indigenous ethnicity in More-los over thirty years ago, residents feel that to a degree "the more material symbols of Hispanic [mestizo] culture they obtain, the less Indian they will become" (1975, 131).

For others, their experience abroad as Mexicans and exposure to other, more positive ideas about Mesoamerican Indians contribute to their rethinking the meaning of being an Indian. In contrast to their parents and grandparents, migrants and maquila workers have access to social prestige through their income and travel experience rather than patron-client agricultural relationships. In this way the experience of working in the United States and in valley maquilas is transforming some notions about community and belonging, while simultaneously reproducing others.

Residents from San José encounter, negotiate, and generate multi-ple and contradictory meanings about what it means to be a campe-sino, an indigenous Mexican, a Sanjosepeño, a maquila worker, or a migrant. Although positive images of indigenous Mexicans and campesinos circulate through the media, in San José, a town region-ally known for its traditional indigenous practices and its history of violence, residents use the term "indio" to highlight violence, de-humanization, and the shared memory of conflict over resources. Some residents discuss their identities as Indians in terms of fierce-ness in the face of injustice. When older Sanjosepeños refer to them-selves as campesinos, as "hard workers," and as being "screwed" by the government—which treats them as inefficient, burdensome producers—they assert their humanity and emphasize the dignified nature of hard work in el campo.

Today in the valley young migrants are often uninterested and unskilled in maize agriculture. While slightly older migrants have more varied work experience and are much more likely to cultivate a milpa, young migrants view agricultural production as an unprofit-able, difficult way of life, practiced by their traditional parents. The question remains whether young migrants will be skilled in maize agriculture as they age, but the option to maintain an agricultural livelihood is being eroded with increasing production costs, declin-ing levels of spring water, and other factors like home construction on arable land. Any debate about the future of Mexican maize and its in situ conservation must address the hardships faced by rural Mexi-cans and their perspectives about life in el campo.

Residents' livelihood strategies under the neoliberal corn regime have paradoxical results. While members of the younger generation face exploitation as disposable, inexpensive workers in transnational labor circuits, their experiences also provide them with a sense of freedom from parental and patriarchal authority. Their involvement in these labor circuits—and their lack of training or experience in agriculture—is experienced as freedom from the peasant-based livelihoods of their parents and grandparents, at least for now. In turn, older residents rely on maize agriculture as a social safety net, but the processes which intensify their recourse to corn also undermine the ability of members of the next generation to remain on the land, if they so choose, as agricultural producers.

This book began with the public controversy about GM corn and then took readers behind the debates to examine the everyday livelihood struggles faced by rural Mexicans. In the southern Tehuacán Valley residents have taken up an intergenerational, transnational household strategy to weather over two decades of local environmental problems, new policies, an influx of cheap corn imports, and the effects of economic crisis. While an older generation relies on the "recourse to corn," younger residents, employed in valley maquilas and in the United States, send part of their income home to help maintain or establish a household in town. Together these different generations maintain, and in some cases advance, the economic standing of the rural household; but this strategy is not without its strains and stresses on individuals and families.

Despite the intentions and expectations of state policy, the Mexican production of white maize increased after NAFTA. My research confirmed what several other studies had found elsewhere in central and southern Mexico—that maize cultivation is subsidized by off-farm income and at the same time provides a safety net for rural producers as a dependable yet flexible crop used for subsistence food and cash income (Barkin 2002; de Janvry, Gordillo de Anda, and Sadoulet 1997; García-Barrios and García Barrios 1994; Nadal 2000b; Warman 2003 [1988]). Yet there has also been an interruption in the transmission of agricultural knowledge from one generation to the next, as young valley residents find work in maquilas and by undertaking transnational migration. And despite the recourse to corn, the difficult conditions for agriculture, the shortage of local agricultural labor, and the lack of interest among young migrants are not the sign of a revitalized countryside but of a deepened struggle to stay afloat. State polices have helped to erode the productive capacity of campesinos and promoted labor migration as rural "development." I agree with Armando Bartra and others that those rural Mexicans who want to work and live as small-scale cultivators should have the ability to do so; they should have the right not to migrate.

I have suggested that these conditions also undermine the in situ diversity and abundance of traditional corn varieties. The transformation of the younger generation of rural Mexicans into labor migrants must be considered when attempts are made to protect the biological diversity of maize. In this sense the southern Tehuacán Valley is representative of wider changes taking place in the Mexican countryside—notably the increase in rural labor migration and the aging of maize farmers—despite the region's reliance on irrigated elote corn.

Changes in the countryside affect valley residents in different and sometimes contradictory ways. As young people join transnational and gendered labor circuits, their journey to work makes them more mobile than their parents or grandparents were in terms of both travel and social prestige. Off-farm wage labor also provides the opportunity for economic advancement, but this is not equally realized among residents. Some individuals and households fare better than others, investing in machinery or water shares, while others are unable to save the funds to do so. On the one hand, young men and women of the valley are disposable and exploited workers, and on the other their work experiences and wages contribute to a sense of social mobility and a loosening of parental and patriarchal authority. In other words, the erosion of peasant livelihoods is paradoxically experienced by the younger generation as freedom from the agricultural livelihoods and parental authority of their elders.

This sense of freedom from travel and a wage comes at a price: it is accompanied by exploitation at the maquila or on the job, and the emotional toll and physical dangers of illegal border crossings and deportability. Even though most young migrants say that they prefer work abroad to agriculture, transnational migration has its social costs. In the words of a returned migrant mentioned in the last chapter, "Migration helps out financially but it tears families apart."

It is paradigmatic of the industrial food system that those from the global south who cannot or prefer not to work the land as the cultivators of food crops now staff restaurant chains and work in food-processing plants in the global north. The Mexican state strategy which encourages the import of corn and the export of labor is symbolic of the growing distance between food producer and consumer in a globalizing, corporate food regime. Although valley residents prefer the taste of local criollo corn, "corn from elsewhere" has made its way to the countryside as grain distributed through government programs, as animal feed, and in the form of industrial tortillas more commonly found in cities.

This book has outlined a few of ways that corn—as both a food and a crop—is fraught with multiple layers of meaning, and how it has been portrayed by state policy, bureaucrats, activists, academics, and local actors. Rural inhabitants and the anti-GM, pro-corn coalition In Defense of Maize negotiate and challenge the policies and framework of the neoliberal corn regime. As part of its campaign against genetically engineered maize and in support of pro-campesino policies, In Defense of Maize has used corn as a symbol of the Mexican nation and culture. Symbolic associations like this have been so successful that the industrial corn flour and tortilla company Gruma professes its support for Mexican culture on its corporate website and declares: "Long live the people of maize!" (Viva el pueblo de maíz).[1]

In their role as policy watchdog—as well as in their forums, publications, and demonstrations—In Defense of Maize advocates food sovereignty and quality over a narrow focus on market efficiency, in the process recasting the question of GM corn imports and field trials in terms of cultural values, peasants' expertise, and corporate-led globalization. This coalition challenges the rule of experts and the positioning of GM regulators, scientists, and state officials as the only legitimate experts in setting the agricultural agenda for rural Mexico. The coalition also demands transparency from state bureaucrats and politicians, and democratic decision making, which takes into account the concerns and expertise of maize cultivators, as well as the concerns of consumers and environmentalists.

In San José residents, migrants, maquila workers, and corn farmers negotiate the multiple layers of meaning and state policy associated with the cultivation of corn and the countryside. Some young Sanjosepeños associate maize agriculture with outdated tradition and indigeneity. But to understand why they view maize production in these terms, this book has explored some of the different social contexts in which workers, migrants, and campesinos find themselves, including the history of struggle over water, the marking of

San José as a particularly Indian town, the travel and wages of young Sanjosepeños, and finally the portrayals of corn as an inefficient crop—portrayals which draw on the earlier associations of corn with backwardness and poverty.

In contrast to the younger generation, I found that older Sanjosepeños describe themselves as campesinos to affirm their dignified yet difficult livelihoods in the face of racist discourses and the neoliberal model. In San José, where a tendency to fight over water and engage in violence was portrayed regionally as an essentially "Indian" trait, the language of ethnicity associated with the indigenous rights movement and recent state support of cultural pluralism have little local appeal. While the neoliberal state employs a language of diversity, it has cut its commitment to redistributing resources and supporting small-scale agriculture—what Charles Hale (2002) has called "neoliberal multiculturalism" in another setting. Given these local and regional contexts and the promotion of neoliberal multiculturalism at the national level, it is not surprising that the campesino identity which resonated with Sanjosepeños during agrarian reform should continue to hold meaning today. While not explicitly articulated as part of a political struggle to retain their productive resources or rights as peasants in making claims on the state, residents' use of the term "campesino" should nevertheless be viewed in light of the economic, political, social, and cultural context of accumulation by dispossession and neoliberal governance. Older residents feel strongly that the state is reneging on its post-revolutionary commitments.

My comparison of the GM maize debates and valley discussions of the issues has highlighted the uneven nature of public debate. Southern valley residents were unfamiliar with the GM controversy brewing just next door in Oaxaca, in Mexico City, and in transnational media and NGO networks. At the same time, despite this dissonance, both valley residents and In Defense of Maize participants have employed the campesino identity and a notion of risk in their critique of the neoliberal corn regime. When In Defense of Maize demands the renegotiation of NAFTA or the inclusion of peasant interests in rural policy, they draw on the post-revolutionary framing of the campesino as an unjustly disenfranchised group of people with economic and political interests in common. In some instances the result has been what has been termed "peasant essentialism"—romanticizing peasants' (not to mention indigenous peoples') experiences and cultural alterity, or overemphasizing the similarity of experience across time and space. However, employing the post-revolutionary category of the campesino has been a useful strategy for including self-identified campesinos in the coalition and linking the issue of transgenic maize regulation to a broader critique of neoliberal policies. This critique has meant that the coalition moves beyond the official focus on the risks of gene flow between transgenic and traditional varieties of maize to incorporate other equally important challenges facing rural producers and in-situ biological diversity, such as trade liberalization, the expansion of corporate agriculture, cuts to rural subsidies, and circular migration and outmigration from the countryside.

As made clear by the narratives of residents of the southern valley, not only does the campesino identity retain powerful political and cultural meaning in the region, but the cultivation of corn is a strategy to guarantee food and maximize household labor. Corn is also seen as a particularly resilient crop. Growing maize is a way to manage the potential loss of income and security that is characteristic of small-scale agriculture and an unreliable labor market. In this sense, without being in direct conversation with each other, In Defense of Maize and Sanjosepeños employ similar notions of what it means to be a campesino and the difficulties facing small-scale maize cultivators.

Since the time I finished my research in the valley, In Defense of Maize has worked to expand its reach and enroll more participants. In 2008 the coalition organized a caravan which started out from the "cradle of corn" in Coxcatlán and traveled through the Tehuacán Valley on its way to Mexico City. The main demands of the caravan were the renegotiation of NAFTA and a halt to any planned tests or imports of GM maize.[2]

My comparison of the GM maize debates and valley discussions of the issues has also rested upon an examination of corn as a commodity and as a cultural practice. The local meanings of maize in San José are evidenced not only in what people say and how they say it—keeping in mind that language can of course be performative—but also in what they do in a nonlinguistic sense. As a commodity purchased, prepared, and consumed in local, regional, and distant settings, maize embodies the social relations of its rural production.

To a large extent, this book has explored how resident livelihoods are affected by the history of state-community interaction by focusing on three key commodities in the valley: water, labor, and corn. Regarding water, I have argued that during the twentieth century irrigation management and conflict profoundly shaped the organization of community as a political administrative unit, a set of unequal social relations, and an imagined sense of shared belonging. The uneven accumulation of water resources also exacerbated water scarcity and culminated in the water conflict of the 1980s. These water shortages and disputes were a factor in pushing residents to seek employment in Mexico City and the United States.

In terms of labor, campesinos in the southern Tehuacán Valley have adapted to the economic and environmental crises of the 1980s by adjusting and altering the labor responsibilities within the household. As suggested by earlier agrarian debates, the adaptability and flexibility of peasant and household labor are central to rural survival. According to Sanjosepeño accounts, some formerly unpaid household tasks have been monetized since the 1970s, particularly in the 1980s. In an area where the division of labor is strongly associated with gender roles, these changes have reproduced some ideas about gender while shifting others.

The experiences of wage labor and migration affect how residents view community, or notions of difference and affinity within and between households. For instance, the mobility of younger residents enables them to feel less Indian than their grandparents and contributes to more positive associations with being an Indian, as do the presence of local role models and an exposure to the images and ideas of the indigenous rights movement. Significantly, however, work in maize production is not considered an activity to emulate.

For some scholars, transnationalism and globalization have provoked new ways of theorizing community relations. Roger Rouse argued that the experience of migration generated a way of looking beyond the definition of "community" as a population contained within a bounded space and consisting of institutional parts working toward the stability of the whole (1991, 10). Similarly, some have proposed a notion of "transnational space" to capture how new migrant social identities and spaces transcend the social structure of both the sending country and the receiving country, crossing cultural, political, and social boundaries (Basch, Blanc Szanton, and Glick Schiller 1995). Transnational migrants are characterized by their continued commitments in both their adopted home and their place of origin, as we saw in the transnational household in San José (Basch, Blanc Szanton, and Glick Schiller 1995, 7). Rouse's call to denaturalize or historicize the space of community emphasized both the constraints imposed by structures and practices of power and their ability to change through the collective engagement of social actors. In this book I have also attempted to avoid notions of a naturalized, internally homogeneous community or culture, in part by historicizing the rural community. One of my arguments has been that long before the recent period of accelerated migration and liberalized trade, the indigenous community of San José was already a flexible labor force shaped by interaction with the wider projects of state and capital. The flexibility of the valley labor force is not new, but the ways that flexibility is manifested, and the extent to which residents are dependent on labor markets, have changed. Valley residents are more connected and vulnerable to fluctuations in the global economy.

Additionally, by focusing on social relations and cultural practices rather than on a shared millennial indigenous or peasant culture, I sought to demonstrate how community resources and agricultural knowledge are not distributed equally throughout the community but rather related to political power, male privilege, and (regarding young migrants' lack of interest in agriculture) the wage labor experience. To avoid simplistic notions of corn culture as bounded communities organized around subsistence, and living in harmony with nature and each other, I focused on the details of local history. Moreover, the grounding of late capitalism and neoliberal policies within their concrete unfolding in a particular place enlarges our understanding of how the forces and processes operate. In the Tehuacán Valley accumulation by dispossession has brought about the further diversification of household strategies and the related remaking of maize agriculture.

Maize is the crop of choice in the southern valley, as it is in many parts of the country, because of its multiple uses and adaptability. Corn is the peasant crop par excellence in Mexico because of its adaptation to a huge range of agro-environments and the flexibility it offers as a dietary staple and a commodity for trade or sale. The cultivation of corn, particularly irrigated elote in San José, helps to cushion the impact of economic crisis on the rural household.

This book has situated valley livelihoods in relation to the corn debates in part to show how the competing sides of the debates portray corn production in places like San José. I wanted to suggest that the obstacles confronted by these seemingly remote maize farmers in the southern valley are affected by (and can inform) "expert" discourses and debates. I also wanted to suggest that these maize cultivators are connected to the food supply of urban consumers in Mexico City, Seattle, and Toronto, as well as the fate of corn farmers in Iowa and Saskatchewan. Foods "from nowhere" exacerbate the already difficult conditions of small-scale agriculturalists; and at the same time, for those of us who consume foods produced, processed, and distributed through the industrial food system, our future food supply may depend in part on small-scale agriculturalists and their maintenance of biodiversity in the field.

CASE 1

Number in household: Five: older couple, adult daughter,
   and two children
Land area: 7 hectares, rainfed and irrigated
Crops grown: Elote, maize, sugarcane
Source of maize seed: Criollo, saved and bought in valley
Milpa labor: Farmer's labor plus *peones* (paid agricultural laborers)
Other income: Prepared food sales, son's office job in city, Procampo
Remittances: Does not receive remittances
Water shares: Yes, plus ravine water

CASE 2

Number in household: Three: couple and one adult son in household
   (plus two married daughters and one son in United States)
Land area: 3 hectares, rainfed and irrigated
Crops grown: Maize, elote, and tomato (*jitomate*)
Source of maize seed: Criollo, bought
Milpa labor: Farmer's labor plus peones
Other income: Goats, farmer's work as peon
Remittances: Receives remittances, spends on agriculture
Water shares: No water shares, but ravine water

CASE 3

Number in household: Two: couple (plus two married daughters in
   town and one son in United States)
Land area: 4 hectares, rainfed and irrigated

Crops grown: Elote, maize, melon, tomatoes, squash, two types of beans
   (one grown with milpa)
Source of maize seed: Local criollo, saved and bought
Milpa labor: Farmer's labor plus peones
Other income: Odd jobs, goats
Remittances: Receives remittances, spends on agriculture
Water shares: Yes

CASE 4

Number in household: Four: couple and two adult children
   (plus two married children)
Land area: 1 hectare rainfed, irrigated if no rain
Crops grown: Elote, maize, beans with milpa
Source of maize seed: Local criollo, saved and bought
Milpa labor: Farmer's labor plus sons
Other income: Children work in maquilas
Remittances: No
Water shares: No

CASE 5

Number in household: Five: couple, two teenaged daughters, one child
   (plus one married son)
Land area: 2 hectares rainfed, water bought if no rain
Crops grown: Maize, tomato, beans with milpa, sugarcane, melon
Source of maize seed: Local criollo, bought
Milpa labor: Farmer's labor plus kin
Other income: Turkeys, odd jobs, eldest daughter in maquila
   (others in school)
Remittances: No
Water shares: No

CASE 6

Number in household: Two: couple (plus one son in United States,
   one in city, three married daughters in town)
Land area: 2 hectares rainfed, some irrigation if no rain
Crops grown: Maize, tomato
Source of maize seed: Local criollo, bought
Milpa labor: Farmer's labor plus peones
Other income: None

Remittances: Receives remittances, spends on agriculture
Water shares: n/a

CASE 7

Number in household: Three: older couple with one returned migrant son
  (plus six other married children, one in United States)
Land area: 5 hectares, rainfed and irrigated
Crops grown: Maize, green beans, tomato, squash
Source of maize seed: Local criollo, bought and saved; tried hybrid
Milpa labor: Farmer's labor plus sons (worked in fields more
  often when younger), peones
Other income: Works as peon, Procampo
Remittances: Receives remittances, spends on agriculture
Water shares: No

CASE 8

Number in household: Four: couple, two young children
Land area: 6 hectares, rainfed and irrigated
Crops grown: Maize, tomato, elote
Source of maize seed: Local, saved and bought
Milpa labor: Farmer's labor plus peones
Other income: Turkeys, Procampo
Remittances: No
Water shares: No, but ravine water

CASE 9

Number in household: Three: couple, teenaged son (plus three
  other married children)
Land area: 1.25 hectares rainfed, irrigated if no rain
Crops grown: Maize, elote
Source of maize seed: Local criollo, saved and bought
Milpa labor: Farmer's labor plus peones
Other income: Procampo, turkeys, chickens
Remittances: No
Water shares: No, but ravine water

CASE 10

Number in household: Six: couple, four children (plus one migrant son)
Land area: 1 hectare

Crops grown: Maize
Source of maize seed: Saved seed
Milpa labor: Farmer's labor plus son, plus peones
Other income: Works as peon, goats, daughter in maquila
Remittances: Yes
Water shares: No

CASE 11

Number in household: Four: couple, one teenaged child, one young child
(plus four sons in United States)
Land area: 12 hectares, rainfed and irrigated (pooled with sons)
Crops grown: maize, elote, melon, watermelon
Source of maize seed: Saved seed
Milpa labor: Farmer's labor, children's help when present, peones,
wife's work with tomatoes
Other income: Small business with son, Procampo
Remittances: Yes
Water shares: Yes

CASE 12

Number in household: Two: older couple (children grown)
Land area: 2 hectares, irrigated
Crops grown: Maize, elote, beans
Source of maize seed: Saved seed
Milpa labor: Farmer's labor plus peones
Other income: Some from adult children
Remittances: No
Water shares: No

CASE 13

Number in household: Nine: couple, one grandparent, six children
of various ages
Land area: 2 hectares, rainfed and irrigated
Crops grown: Maize, elote, beans
Source of maize seed: Local, saved; bought if needed
Milpa labor: Farmer's labor plus sons, and *a medias* (sharecropping)
Other income: Procampo, father and sons sharecrop and work as peones
Remittances: No
Water shares: No, but ravine water

CASE 14

Number in household: Four: couple, two young sons
Land area: 3.5 hectares rainfed, some irrigation
Crops grown: Maize, elote, squash and beans with milpa
Source of maize seed: Saved, rarely bought
Milpa labor: Farmer's labor plus peones
Other income: Small furniture workshop
Remittances: No
Water shares: No

CASE 15

Number in household: Four: couple living part-time in Tehuacán with
   children
Land area: 2 hectares, irrigated
Crops grown: Maize, elote
Source of maize seed: Bought seed
Milpa labor: Farmer's labor plus peones
Other income: Works as plumber
Remittances: No
Water shares: No, but ravine water

CASE 16

Number in household: Four: widow, adult daughter, two small grandchildren
   (plus two sons in United States)
Land area: 1 hectare, irrigated
Crops grown: Maize, elote
Source of maize seed: Local seed, bought
Milpa labor: Peones only
Other income: Daughter in maquila, widow works in corner store,
   and piecework embroidery
Remittances: Yes
Water shares: No

CASE 17

Number in household: Three: older couple, parent
Land area: n/a
Crops grown: Maize
Source of maize seed: Saved seed
Milpa labor: Farmer's labor a medias plus peones

Other income: Husband works as peon, wife embroiders and
    sells fruit at market
Remittances: No
Water shares: No

CASE 18

Number in household: Six: couple, four children of various ages
    (plus three adult children who are migrants with own households)
Land area: 4 hectares rainfed, but often needs to buy water
Crops grown: Maize, elote, sugarcane
Source of maize seed: Saved and bought
Milpa labor: Farmer's labor plus sons
Other income: Works as peon, two children in maquilas, two sons
    work sporadically as migrants
Remittances: Yes
Water shares: No

CASE 19

Number in household: Two: older couple (three adult children
    were all migrants)
Land area: Area not available, rainfed (and buys irrigation water)
Crops grown: Maize, elote
Source of maize seed: Saved
Milpa labor: Farmer's labor plus peones
Other income: Procampo, turkeys, occasional funds from children
Remittances: No
Water shares: No (but had formerly, before water dried up)

CASE 20

Number in household: Four: couple, two school-age children
    (all brothers are in United States and return home to farm)
Land area: 2 hectares plus rented land, rainfed and irrigated
Crops grown: Maize, elote
Source of maize seed: Saved
Milpa labor: Famer's labor plus brothers
Other income: Works as peon; wife raises pigs and turkeys, sells
    prepared food, and embroiders
Remittances: No
Water shares: No

CASE 21

Number in household: Three: older couple, grandparent
  (plus six sons and two daughters in own households)
Land area: 6 hectares, rainfed and irrigated
Crops grown: Maize, elote, zucchini, tomato
Source of maize seed: Saved
Milpa labor: Farmer's labor plus peones, sons (in their thirties)
  when back from United States
Other income: Pigs, goats, chickens, occasional help from children,
  wife's embroidery
Remittances: No (but had formerly, when migrants did not have
  their own families)
Water shares: No, but ravine water

CASE 22

Number in household: Four: mother, three school-age children
  (husband in United States)
Land area: 1.5 hectares rainfed (buys water when needed)
Crops grown: Maize, elote
Source of maize seed: From father-in-law
Milpa labor: Hires peones
Other income: Turkeys, remittances, wife's embroidery
Remittances: Receives remittances, spends on agriculture
Water shares: No

CASE 23

Number in household: Two: widower, daughter (plus daughter's
  long-term migrant husband)
Land area: 1.5–2 hectares, rainfed (buys water when needed)
Crops grown: Maize, elote
Source of maize seed: n/a
Milpa labor: Farmer's labor plus peones
Other income: Farmer works as peon, wife's embroidery
Remittances: Occasional
Water shares: No

CASE 24

Number in household: Six: couple, four children in school
  (husband is former migrant)

Land area: 1 hectare borrowed, buys water

Crops grown: Maize, elote, tomato, zucchini, melon

Source of maize seed: Seed bought in valley town

Milpa labor: Farmer's labor plus peones, wife works with jitomate and melon

Other income: Farmer now works in mayor's office

Remittances: No (but saved money as migrant)

Water shares: No

CASE 25

Number in household: Four: parents, former migrant daughter and her baby

Land area: 5 hectares, rainfed and irrigated

Crops grown: Maize, elote

Source of maize seed: Saved

Milpa labor: Farmer's labor plus sons (who live in their own homes)

Other income: Small store, sales of tortillas, embroidery, remittances from daughter's husband

Remittances: Yes

Water shares: No (farmer had formerly, before water dried up)

CASE 26

Number in household: Five: elderly couple, daughter (plus husband in United States long-term), young children

Land area: 3 hectares, rainfed and irrigated

Crops grown: Maize, elote (grew more until 1999 but can no longer afford to do so—now grows only enough to eat)

Source of maize seed: Saved

Milpa labor: Farmer's labor plus peones (formerly practiced a medias)

Other income: Remittances, embroidery, turkeys

Remittances: Receives remittances, spends on agriculture

Water shares: No

CASE 27

Number in household: Five: couple (husband migrates to United States), young child, grandparents

Land area: n/a

Crops grown: Maize, elote

Source of maize seed: Saved

Milpa labor: Grandfather's labor plus family

Other income: Grandfather works as peon, husband migrates, embroidery

Remittances: Receives remittances, spends on agriculture
Water shares: No

CASE 28

Number in household: Two: young couple (he is returned migrant)
Land area: Works on parents' land
Crops grown: Maize, elote, tomato
Source of maize seed: Saved
Milpa labor: Farmer's labor plus family
Other income: Returned migrant, works as well digger
Remittances: Does not receive remittances (did formerly and
    spent on agriculture)
Water shares: No

CASE 29

Number in household: Four: couple, two children in school
Land area: 3 hectares, irrigated
Crops grown: Elote
Source of maize seed: Criollo, bought in valley town
Milpa labor: Farmer's labor, a medias
Other income: Food sales, Procampo
Remittances: No
Water shares: Yes

CASE 30

Number in household: Two: couple (plus two adult children in
    United States and five married children)
Land area: Area not available, buys water from Chilac
Crops grown: Maize, elote
Source of maize seed: Saved
Milpa labor: n/a
Other income: Farmer works as peon, remittances from children
Remittances: Yes
Water shares: No

INTRODUCTION: The Struggle for Mexican Maize

1 In this book, I follow common usage by employing the terms "genetically modified," "genetically engineered," and "transgenic" interchangeably. Some scholars prefer the terms "genetically engineered"
and "transgenic" because plant breeding and non-biotech agricultural
crops involve the modification of genes without the use of genetic
engineering.

2 This includes the production of certified organic fruits and vegetables.
Certified organic has risen to 2.3 percent of total production in Mexico,
90 percent of which is for export. "Mexico Boasts the Highest Number
of Organic Farms," Greenplanet.net, 10 April 2009.

3 The National Ecology Institute (INE) and the National Commission on
Biodiversity (CONABIO) sampled maize ears and harvested grain from
twenty-one locations, including two from the northern Tehuacán Valley
in the Tehuacán-Cuicatlán region, where they found evidence of transgenic introgression (INE-CONABIO 2002; Ezcurra, Ortiz, and Soberón 2002, 280).

4 Regino Melchor Jiménez Escamilla, "Estadísticas de San José Miahuatlán, Puebla," 25 August 2005 (unpublished local survey).

5 The PRI was founded in 1929 as the Partido Nacional Revolucionario.
It was renamed in 1938 as the Partido de la Revolución Mexicana and
again in 1946, when it was given its current name.

6 The PRD was founded in 1989 by disgruntled members of the PRI and
other leftist politicians after Cuauhtémoc Cárdenas, the candidate of a
center-left coalition and son of the famous president of the 1930s, was
denied the presidency through electoral fraud in 1988. The PRI candidate, Carlos Salinas, was declared president. The PRD was formed by a
coalition of smaller left-wing parties such as the Unified Socialist Party
of Mexico (PSUM). The PSUM itself had also been formed through a coalition of leftist parties, including the Mexican Communist Party (PCM).
In the late 1980s PSUM changed its name to the Partido Mexicano
Socialista (PMS) before joining other leftist parties to form the PRD.

7 I refer to San José by its real name rather than a pseudonym because readers familiar with the valley can easily ascertain the town's identity based on its location, traditions (like marriage brokers), and history of water disputes. Additionally, I refer to the *tetlallí* Carlos Vargas by his real name (with his permission) because he was a respected elder knowledgeable in local history. A tetlallí is a marriage broker (discussed further in chapter 3). I also refer to the teacher Regino Melchor Jiménez Escamilla by name, with his permission. Regino and I conducted several interviews together in which he translated Nahuatl to Spanish. Finally, I have not changed those residents' names that were published in newspaper articles or land and water title documents from the 1980s or earlier.

8 Tom Brass (1997) and Henry Bernstein (2006) also use the term.

9 "NAFTA Commission Gets GM Corn Complaint," *Frontera NorteSur*, 28 January 2009.

10 Thanks to David Barkin for helping to clarify this point.

11 The concept is also a lens for historicizing the political economy of food, trade, and agriculture (McMichael 2009).

12 The National Action Party (PAN) was founded by conservative Catholics in 1939 as an opposition party. Today it is a conservative-populist party.

13 This is a 3.6 percent drop from the previous year, believed to be caused by the global financial crisis and the crackdown by the United States on undocumented immigration. Elisabeth Malkin, "Money Sent Home by Mexican Workers in U.S. Falls Sharply," *New York Times*, 2 June 2009.

14 The Bretton Woods accord came out of a United Nations Monetary and Financial conference held in 1944 at which countries of the global north met to discuss new economic policies to help promote reconstruction in Europe after the Second World War. These policies included tying currencies to gold reserves (that is, establishing the international gold standard exchange rate system), reversing trade protectionism, and establishing the General Agreement on Tariffs and Trade (GATT), the International Monetary Fund (IMF), and the International Bank for Reconstruction and Development, later known as the World Bank.

15 Kopytoff points out that part of the commodity's value comes from the process of its use and not just the social relations of its production (1986, 83).

16 See the edited collection by Stone, Haugerud, and Little (2000) and Barber's and Lem's special issue of *Anthropologica* (2004). The innovative study by Sidney Mintz (1985) of sugar traces its history of production and some of its symbolic meanings and consumption practices. Arjun Appadurai's exposition on the social life of commodities (1986) argues that the studies of such economic goods should take seriously the ways cultural frameworks and power shape demand and consumption, rather than focus exclusively on production. Although he empha-

sizes consumption, Appadurai proposes an approach, based on the work of Kopytoff, which follows the total trajectories of commodities as they move in and out of the commodity stage, from production through exchange and consumption.

CHAPTER 1: Transgenic Maize and Its Experts

1 For example, see the letter defending requests to plant and grow transgenic maize in Mexico by Juan Pablo Ricardo Martínez-Soriano and Diana Sara Leal-Klevezas, published in *Science* 287 (5457), 1399, entitled "Transgenic Maize in Mexico: No Need for Concern." A response was published by Ronald Nigh, Charles Benbrook, Stephen Brush, Luis García Barrios, Rafael Ortega-Paczka, and Hugo R. Perales in *Science* 287 (5460), 1927.

2 This slogan was the title of an exhibit at the Museum of Popular Culture in Mexico City and of an edited collection (Esteva and Marielle eds. 2003). It has been widely used by the anti-GM coalition.

3 Personal communication with Alejandro Espinosa Calderón, investigador titular, INIFAP and UNAM. 18 July 2008 and 19 May 2009.

4 In addition, "terminator" seed—or "genetic use restriction technologies" (GURTs)—also overcomes the biological barrier to commodification. In 1998 the United States granted a patent for a genetic mechanism which produces plants with sterile seeds. Sterility protects seed companies in places where intellectual property rights are not recognized. The biotech industry has recently argued, however, that such technology is needed to prevent the escape of transgenes into the environment. This seed was such a public relations nightmare that to date it has not been marketed. RAFI, now the ETC Group, coined the term "terminator" to refer to the technology (Stone 2002, 613; see also McAfee 2003b).

5 Rural auditors had taken samples from Schmeiser's property in Saskatchewan without his permission. Schmeiser contested Monsanto's claim by arguing that he had not knowingly planted GM seed and had been an innocent bystander. When the case was taken to the Canadian Supreme Court the court ruled in favor of Monsanto but did not require Schmeiser to pay Monsanto its court fees nor the profit made from the crop. Instead he was ordered to turn over to Monsanto any remaining canola plants or seed which contained the patented material (Müller 2006a). When Mexico began legislative hearings on a proposed biosafety law (LBOGM), Schmeiser was invited to speak about his experience with Monsanto.

6 Cleveland and Soleri (2005, 6) outline four main variables and then discuss possible scenarios: "1) the volume and frequency (unique or sporadic) of pollen flow (m, migration); 2) fitness (W) including rela-

tive fitness of different genotypes and absolute (or ecological) fitness of populations; 3) selective pressure (*s*) exerted by biophysical and social environments, including both conscious and unconscious farmer selection; and 4) changes in the selection pressures, such as the evolution of resistance in local pest population or climate change." See also Bellon and Berthaud, 2006.

7 For a summary of the various studies on Bt corn and other crops see Thies and Devare 2007.

8 In 2002 Mexico signed and ratified the Cartagena Protocol, which had been adopted in January 2000.

9 The Precautionary Principle was first employed internationally in the World Charter for Nature, adopted by the UN General Assembly in 1982. It has been used in various international treaties and declarations, such as the Montreal Protocol (1987) and the Rio Declaration on Environment and Development (1992). In the 1990s the United States and the EU disputed the scientific basis of the principle in risk assessment. As the National Research Council of the National Academies (2002, 65a) points out, there are multiple understandings of this principle, ranging "from minor procedural changes in risk analysis to major shifts in burden of proof."

10 The trilateral "Documentation Requirements for Living Modified Organisms for Food or Feed or for Processing," was signed by Víctor Villalobos, coordinator of international affairs for the Ministry of Agriculture (SAGARPA).

11 The agreement also makes the labeling of such imports available only to distributors and not consumers. Additionally, critics have argued that this trilateral agreement has no legal basis, because it was signed without being presented to the Mexican Senate (Ribeiro 2003).

12 See Worthy, Strohman, Billings, Delborne, Duarte-Trattner, Gove, Lathan, and Manahan (2005) for a discussion of how the work was presented as sloppy and biased. Delborne (2005) also shows how Chapela's and Quist's claims about gene flow were portrayed as the sloppy use and interpretation of iPCR data. Some critics of the study argued that using southern blots rather than the PCR, which is prone to false positives, would be more reliable.

13 For example the minister of agriculture's statement, mentioned in "Exigen organizaciones que México exponga el caso de la contaminación del maíz en la Cumbre Mundial de la Alimentación," 10 June 2002, received from ETC Group and maiceros-l@laneta.apc.org. In a newspaper interview a representative of Monsanto said, "The fact is that the biotech traits really don't pose any unique risk to the local maize" (Paul Elias, "Corn Study Spurs Debate over Corporate Meddling in Academia," Associated Press, 18 April 2002). A scientists' statement on the *Nature* controversy, made available by AgBioWorld, says that "the kind of gene flow alleged in the *Nature* paper is both inevitable and wel-

come" ("Joint Statement in Support of Scientific Discourse in the Mexican GM Maize Scandal," 24 February 2002, http://www.agbioworld.org/jointstatement.html).

14 In the article "Exigen organizaciones" cited in note 13, above.

15 In the early 1990s Dr. Herrera Estrella, a molecular biologist, was involved in the development of a virus-resistant potato at CINVESTAV using technology donated by Monsanto, a world leader in agricultural biotechnology. Monsanto granted the center a royalty-free license to use its technology. CINVESTAV is a public research institution which also carries out rDNA studies on plants under contract to producers' associations. More recently Dr. Herrera Estrella headed up Proyecto Maestro de Máiz, the partnership between public research institutes in Mexico and several large agro-corporations to conduct tests on GM corn (discussed at the end of the chapter).

16 Although no scientific data proves that GMOs are harmful to human health, the Mexican Academy of Sciences (AMC) maintains that there is no absence of risk in their use. Angélica Enciso, article in La Jornada, 22 April 2003.

17 Greenpeace Mexico sent samples of Maseca and Minsa flour used in making tortillas from several states to laboratories for PCR testing in 2005 and found positive results. Greenpeace Mexico, Boletín 0702, 30 January 2007.

18 Starlink contains a protein, Cry9C, not found in other Bt corn. It is safe for animals but "may trigger allergic reactions in humans, including fever, rashes or diarrhea, according to government scientists." Marc Kaufman, "Corn Woes Prompt Kellogg to Shut Down Plant," Washington Post, 21 October 2000. See also Kaufman's "Biotech Corn Fuels Recall," Washington Post, 23 September 2000, and Schurman 2003.

19 Using visual aids, the authors and their colleagues presented the farmers with two varieties, not identified by name, that represented a local variety (farmers' variety) and a transgenic variety. The transgenic variety represented "a series of hypothetical varieties with properties of hypothetical, locally appropriate Bt maize varieties, and much higher yields than [the local variety in] the first years due to lower pest damage. However, as a result of the evolution of pest resistance to the Bt transgene, yields fall, and to regain high yields, one [transgenic] variety has to be periodically replaced by a new one purchased from the formal seed distribution system" (Soleri, Cleveland, Aragón, Fuentes, Ríos, and Sweeney 2005, 160).

20 A month before the release of the CEC report in 2004, Michelle Chauvet and Jorge Larson, a sociologist and a biologist, conducted five workshops in the sierra norte of Oaxaca summarizing the scientific findings of the CEC report (Larson and Chauvet 2004; interview with Jorge Larson, 28 July 2006). They led workshops for farmers in Oaxaca on what the study showed, what GM corn is, and how best to save criollos.

21 Such as PCR, which refers to polymerase chain reaction, a technique used to amplify DNA exponentially.

22 Several Mexican organizations—GEA, Greenpeace, ANEC, CECCAM, UNORCA—and the economist Alejandro Nadal submitted a complaint to the Federal Environmental Protection Agency (PROFEPA) in the late 1990s about a bioprospecting contract to collect and export plant samples between a biotech firm from California, Diversa Corporation, and the National Autonomous University of Mexico (UNAM). The contract was canceled, along with the ICBG-Maya agreement for the southern state of Chiapas (interview with Catherine Marielle, 27 July 2006). A forum in Mexico City entitled "Bioprospecting or Biopiracy? Biodiversity and the Rights of Peasants and Indigenous Peoples" was organized in the fall of 2000 by CECCAM, CASIFOP, and RAFI (now the ETC Group) to strategize around the issue. For a thorough discussion of bioprospecting in Mexico see the work of Cori Hayden on the agreement between UNAM and the Institute for Collaborative Biotechnologies (ICB) based at the University of Arizona (2003a; 2003b).

23 Rather than address the issue of GM corn in detail, the agreement states: "Measures to protect agro-biodiversity will be implemented, such as monitoring contamination and genetic erosion." The National Peasant Confederation (CNC) accepted the proposal. Available at http://www.inca.gob.mx/archivos_source/acuerdo%20completo .pdf.

24 In "En defensa del maíz y contra la contaminación transgénica," news release issued by civil society organizations (CASIFOP, CECCAM, ETC Group, ANEC, CENAMI, COMPITCH, FDCCH, FZLN, Greenpeace, Instituto Maya, SER Mixe, UNORCA, UNOSJO, and RMALC) in Mexico City on World Food Day, 16 October 2001. Translated by ETC Group.

25 Other examples include "Of the People of Corn," meeting in Oaxaca in 2003 organized by CENAMI (National Support Center for Indigenous Missions); "Peasant Encounter on Seeds and Life," organized in Mexico City in 2004 by UNORCA; and Strategic Community Workshops on the Contamination of Native Corn, organized in late 2005 by CECCAM, at which peasants from various states exchanged photos and test results on transgenic corn from their fields. The Ministry of the Environment published a cartoon-based leaflet for distribution in the countryside entitled, "How to Conserve Our Native Corn."

26 Under the direction of the National Council for Culture and the Arts (CONACULTA), a museum exhibit entitled "Sin Maíz, No Hay País" opened at the National Museum of Popular Culture in Mexico City. The group also edited and published a book of the same name (Esteva and Marielle eds. 2003). As part of the World Day of Action against GMOs in April 2006, GEA organized the first non-GM food trade fair, "Primera Feria por una Alimentación Libre de Transgénicos," in Mexico City, Oaxaca, Texcoco, Tlaxcala, Guadalajara, Puebla, and Uruapan,

along with several other organizations. At the fair they launched a logo that indicates when a product is GMO-free.

27 See "Defender nuestro maíz, cuidar la vida," *GRAIN* 40 (April 2004) online at http://www.grain.org/biodiversidad/?id=227. See also María Colín, "Bioseguridad en México: una cronología," *La Jornada*, 13 May 2008, http://www.jornada.unam.mx/2008/05/13/amenaza.html.

28 "NAFTA Commission Gets GM Corn Complaint," *Frontera NorteSur*, 28 January 2009.

29 Matilde Pérez, "Sagarpa anula permiso de siembra experimental de maíz transgénico," *La Jornada*, 28 November 2005. Monsanto planned to cultivate 4,352 square meters of a corn resistant to insects and tolerant to a herbicide. The spokesperson for SAGARPA, José Luis Luege Tamargo, said that "they will not permit any violation of the law." Angélica Enciso, "Greenpeace: se ignoró a la Cibiogem al permitir los cultivos de maíz transgénico," *La Jornada*, 17 November 2005.

30 In the following year the corporations of Monsanto, Pioneer-DuPont, and Dow again submitted requests as part of the Proyecto Maestro de Maíz to experiment with seven types of transgenic corn in the fields of public research institutions located in northern Mexico. But the office responsible for ruling on such requests (SENASICA) could not yet grant permits to conduct experimental trials because neither the special regimen nor the country's centers of maize biological diversity had been established; the proposed field trials were suspended (Angélica Enciso, "Evidentes contradicciones entre las secretarías de agricultura y de medio ambiente," *La Jornada*, 25 September 2006). Also controversial was the Ministry of Agriculture's plan to give 4.9 million pesos (around $470,000) to the project, even though the trials involved corporate seed technology and would thus benefit these corporations. One INIFAP researcher, Alejandro Espinosa, questioned SAGARPA's decision to help fund the testing of industry seed in his written comments to SENASICA about the project (Angélica Enciso, "Denuncia experto que Sagarpa apoya a corporaciones de transgénicos," *La Jornada*, 21 February 2006).

31 The manifesto, signed by Mexican scientists, representatives of NGOs, and academics (as well as the author), was circulated and signed by others internationally and published in several Mexican newspapers.

32 Angélica Enciso, "Determinan Semarnat y Sagarpa que Sinaloa, Sonora y Tamaulipas no son centros de origen," *La Jornada*, 10 November 2006.

33 Angélica Enciso, "Cancelan empresas proyecto para sembrar maíz transgénico," *La Jornada*, 13 March 2008.

34 "Autorizan siembras de maíz transgénico experimentales," *Notimex*, 15 October 2009; J. Sanchez, "Permiten el uso experimental de maíz transgénico en México," *El Universal* (national newspaper), 7 March 2009; Alejandro Nadal, "Maíz transgénico, funcionarios delincuentes," *La Jornada*, 11 March 2009.

CHAPTER 2: Corn and the Hybrid Nation

1 The *ejido* is partially based on the *calpulli*, an Aztec unit of social and political organization which was smaller than a city-state and made up of residents who were collectively responsible for political and religious tasks. Residents had access to land through the calpulli.

2 One attempt to bring the Green Revolution to smallholder producers was initiated in 1967. CIMMYT, the International Corn and Wheat Institute, implemented the Plan Puebla to increase yields among smallholder rain-fed maize farmers in the Puebla valley, outside the capital city. The area had been the site of land invasions in previous years (Pansters 1990, 92). In 1974 the Ministry of Agriculture took over the project and expanded it to the sierra and to some other areas. It met with mixed results (Lewontin 1983; Plan Puebla 1974).

3 There are numerous accounts of this debate. I draw on Bartra 1974, Cook and Binford 1990, de la Peña 1988, Hewitt de Alcántara 1984, Kearney 1996, Otero 1999, Redclift 1980, Roseberry 1993, and Stavenhagen 1969.

4 In late 2002 and early 2003 the coalition El Campo No Aguanta Más (the countryside can't take it any more!) organized a week of demonstrations in which half a million protesters demanded the renegotiation of NAFTA and a ban on GM corn, among other things, as well as symbolic border closings at airports, ports, and border areas. The position against transgenic crops was one of six demands made by UNORCA (among other groups) in their call for the December protests ("Movilizaciones campesinas," e-mail message from unorcalistas@laneta.apc.org, 29 November 2002). The organizations involved were the Asociación Mexicana de Uniones de Crédito del Sector Social, Asociación Nacional de Empresas Comercializadoras de Productores del Campo (ANEC), Coordinadora Estatal de Productores Cafetaleros, Coalición de Organizaciones Democráticas Urbanas y Campesinas, Central Independiente de Obreros Agrícolas y Campesinos, Coordinadora Nacional de Organizaciones Cafetaleras, Coordinadora Nacional Plan de Ayala, Frente Democrático Campesino de Chihuahua, Frente Nacional en Defensa del Campo Mexicano, Red Mexicana de Organizaciones Campesinas Forestales, Unión Nacional de Organizaciones en Forestería Comunitaria, Unión Nacional de Trabajadores, El Barzón Nacional, and Unión Nacional de Organizaciones Regionales Campesinas Autónomas (UNORCA). The CNC was initially part of the protests but broke off its affiliation with El Campo No Aguanta Más in January. The leftist press argued that the Ministry of Agriculture (SAGARPA) was trying to divide the movement by negotiating a national agreement on agriculture with PRI member organizations like the CNC ("El Campo ante de la TLCAN," *La Jornada*, 30 January 2003).

5   The complete document can be consulted at http://www.ceccam.org .mx/ConclusionesDefensa.htm.

6   For a complete document see http://www.endefensadelmaiz.org.

7   Bangalore Declaration of the Via Campesina, 6 October 2000, cited in McMichael 2006, available at http://www.viacampesina.org/en/.

8   See website at http://www.slowfood.com.

9   Owner, Amado Ramírez Leyva, quoted in "Tortillería Preserves Local Traditions" on the CIMMYT website, http://www.cimmyt.org, accessed 11 September 2003. For analysis of the Itanoní Tortillería as part of an alternative food network see Baker 2007.

CHAPTER 3: Community and Conflict

1   Despite the implementation of a restricted construction zone in 1946 by the Ministry of Water Resources because of concerns that new *galerías* and deep wells would lower water flow rates, eighty-eight new galerías were built from 1944 to 1969 (Enge and Whiteford 1989, 38). Importantly, the implementation of the *zona de veda* signaled a new level of state intervention in regional irrigation: although associations are private and locally managed, they must receive state permission for construction and cleaning.

2   For example, the Coyoatl spring used by Sanjosepeños went from 43.6 lps in 1976 to 39 in 1983 to 36 in 2002, according to interviews with socios. Late in 2002 the rate went up to 60 lps, which socios believe was due to the closing of two deep wells in Altepexi. See the excellent ethnography on irrigation in the valley by Enge and Whiteford, particularly pages 50–54 on declining flow rates.

3   Galerías have been the main source of irrigation water in the valley since the 1980s (Enge and Whiteford 1989, 36, citing SARH report, *Estudio socioeconómico y cultural corredor Poblano-Oaxaqueño region de galerías Valle de Tehuacán, Puebla* (1982), 34).

4   The information on barrios and mayordomías in San José is based on conversations with Regino Melchor Jiménez Escamilla, the tetlallís Don Carlos and Don Juan, and several other elders between 2001 and 2008.

5   In Regino Melchor Jiménez Escamilla's survey (2002), among the population over the age of fifteen (7,090), 28 percent were illiterate, 29 percent did not complete primary school, and an additional 25 percent did not finish middle school ("Población de 15 años en Rezago Educativo por Municipio," unpublished report, 12 December 2002).

6   The mayor's office is for both the town and the county of San José. There are thirty positions in the *ayuntamiento*. The mayor (*presidente del municipio*) is one of eight council members (*regidores*). There are twenty-

two other positions, which include secretaries, the comptroller, and the police chief and his six police officers.

7 Gene Wilken, *Studies of Traditional Resource Management in Traditional Middle American Farming Systems* (1979), cited in Enge and Whiteford 1989, 31.

8 From the *Diario Oficial* (Federal Register), 11 February 1947, 13, AGN, and *Diario Oficial*, 11 February 1947, Dept. Agrario #34, AGN.

9 The newspaper articles I saw from this period of agrarian reform (1920s–1940s) portrayed political factionalism and violent conflict in racialized terms ("Crónica de Sangre y Escándalo," 20 December 1925; "Destacamentos en Chilac y Miahuatlán," 10 January 1926; "Los sangrientos sucesos de San José Miahuatlán," 7 March 1926). Here is a typical passage from the newspaper *Iris*, based in Tehuacán (17 May 1931; my translation): "During the past two weeks cadavers and injured people from various towns of the ex-district have arrived at the hospital and graveyard killed with the real brutality of savages for questions of little importance because nothing justifies such cruel behavior . . . Lately, our Indians seem only to dedicate themselves to murder, and are losing, little by little, a notion of morality and respect for the life of their fellow man [*prójimo*]."

10 In 1931 the government acknowledged by publishing an announcement in the *Diario Oficial* that the residents of San José had requested a grant, or *dotación*, of water from the two springs (*Diario Oficial*, 25 March 1931, 348, AGN; Atzompa in *Diario Oficial*, 7 March 1931, section 1 (4), no. 17; Coyoatl, *Diario Oficial*, vol. 62, no. 39). See also "Título de confirmación de derechos otorgado al señor Joaquín Albuerne, depositario administrador de la hacienda e ingenio de San Francisco Javier Calipan . . . ," *Diario Oficial*, 13 June 1935, 567, AGN. A request was submitted in 1927 and passed on to the Ministry of Agriculture and Development, the agency responsible for water access at the time.

11 Access to land and water were not always linked in practice. In the same year, 1936, the auxiliary town of San Jerónimo Axochitlán in the municipio of San José also had its petition for water from La Ciénaga and Coyoatl to irrigate land denied, even though it had successfully petitioned and received an *ejido* (*dotación* of 560.4 hectares) from the haciendas of Calipan and Axusco in 1930. The state governor declined the request for water because granting it would reduce the area planted with sugarcane for the nearby Calipan sugar refinery that was using the water (*Diario Oficial*, 23 September 1936, 8, AGN).

12 It may have been the early 1950s, since accounts differ. Nicolás Vargas, the father of the tetlallí, and Francisco Jiménez were among the group that went to Mexico City to submit the paperwork for Coyoatl.

13 While there are 117 water shares listed in the original water association document, *Padrón de usuarios de aguas del manantial "El coyoatl"* (1954),

there are 121 socios listed. Women socios are listed among the original founders of the sociedad, but it was explained to me that there were only 80 socios in the beginning but 117 *acciones* (each acción is 6 hours), so the group wrote down the names of family members, including female relatives, to complete or take up the total number of acciones. In the 1950s there were many more male residents in town than there were acciones, so it seems that the point here was to distribute water shares among friends and allies. Unlike in other valley towns where women are sometimes important water socias, this has not been the case in San José.

14 *Diario Oficial*, 11 February 1947, 13, AGN.

15 "Resolución en el expediente de dotación de aguas al poblado de San José Miahuatlán, Estado de Puebla," 4 June 1936, *Diario Oficial*, 11–12, AGN.

16 The association is called the Sociedad Explotadora y Distribuidora de las Aguas "Agua Grande" y "San Agustín." Although the association was formed in 1936–37 when the water was purchased, there had been an earlier association of the Atzompa water in San José called San Agustín. This information is based on the record of sale kept by the Association, dated 8 May 1937, *Notaría pública de venta de las aguas "Agua Grande" y "San Agustín" de Sra. González a Mucio Galicia, presidente y representante de sociedad civil por acciones "La Sociedad Explotadora y Distribuidora de la Aguas 'Agua Grande' y 'San Agustín.'"*

17 Each month the ejido of Chilac kept eleven days and sixteen hours of water from San Agustín of Atzompa (Enge and Whiteford 1989, 104).

18 Meanwhile, in 1957 the second request for a dotación of water from Atzompa was denied by the governor of Puebla. All the possible sources of water from Atzompa had already been distributed to the nearby valley towns of Coxcatlán, Calipan, San Sebastián Zinacatepec, and San Gabriel Chilac (*Diario Oficial*, 17 October 1958, AGN).

19 In the 1970s several leftist parties, including the Mexican Communist Party (PCM), formed a coalition. These parties then merged to become the PSUM in 1981. Later in the decade the PSUM changed its name to PMS, the Partido Mexicano Socialista, before joining other leftist parties to form the PRD. At the level of local politics in San José, the predecessor to the PRD was PSUM.

20 Three other people survived the same attack. Information is from the regional newspaper *La Escoba*, 6 November 1981, PNA.

21 "San José Miahuatlán sigue siendo un pueblo sin ley," *La Escoba*, 8 July 1983. An official of the *ayuntamiento*, Basilio R. Lezama, was murdered in July 1983 (*La Escoba*, 6 July 1983, PNA).

22 President Salinas (1989–94) put the new National Water Commission (CAN) in charge of redesigning the system for both urban and rural areas (Whiteford, Bernal, Díaz Cisneros, and Valtierra-Pacheco 1998, 384). For research on the effects of more recent neoliberal policy on

water management in Mexico see Barkin 2006; Buechler 2005; Whiteford and Bernal 1996; and Whiteford and Melville eds. 2002.

23 *Diario Oficial*, 11 February 1947, AGN.

CHAPTER 4: Remaking the Countryside

1 Barkin cited in Kay 2008, 931 n. 19.

2 Because elote in the valley has a relatively low yield, the valley is sometimes surpassed in the quantity of elote harvested by those states with
higher yields. See tables 3 and 4.

3 The Coyoatl committee opened the local CONASUPO store. One member recalls, "There was no maize for a couple years. The crops had a
'plague.' So we had to buy yellow corn to eat from Conasupo. It wasn't
very good." Although there were plans to have CONASUPO purchase
local corn, it never did (interviews, 27 November 2001 and 31 January
2002).

4 While these statistics and resident accounts point to general trends in
corn cultivation, interestingly SAGARPA officials expressed contradictory opinions about whether corn production was stable or on the
decline in recent years (interviews, 9 May 2001 and 25 February 2002).
Residents insist that agriculture in general is on the decline, noting that
fields are being converted into houses and that the majority of young
men are engaged in off-farm employment. Yet 2003 proved a good year
for sugarcane and may signal a growing preference for cultivating
sugarcane in the future.

5 In San José ejido land is located in the more marginal and rockier
terrain of the hillsides, and is only occasionally grown with low yield-
ing rain-fed corn. Irrigated land is either communal or owned privately.
Residents with privately owned land near San Pedro rotate sugarcane
production and corn cultivation, both of which are irrigated. Those
with land just beyond the ravine grow corn (including elote) in what
they classify as "muddy soil" all year long because they have access to
the "wild water" of the ravine. Those with communal land cultivate
irrigated elote and rain-fed corn.

6 Sanjosepeños report five irrigations in total. This description of the
agricultural cycle is based on the account by Enge and Whiteford (1989)
and my fieldwork conversations and observations (2001–2, 2005).

7 This was 2,628.50 pesos. The exchange was calculated based on a value
of $0.10316 for 28 February 2001. For mid-March 2001 the rate had
changed to $0.10426. All rates calculated at: http://www.bankofcan
ada.ca/en/rates/exchform.html and verified with Federal Reserve Sta-
tistical Release, www.federalreserve.gov/releases/H10/Hist/datoo_Mx
.txt, accessed 11 September 2003.

8 Gerardo Otero has estimated that four hectares of corn are necessary

for a household of five and a half members, but this seems high (1999, 60). In the sierra de Puebla, Pierre Beaucage found that one hectare of land with fertilizer (or two hectares of unfertilized land) was enough to feed a household of the same size. However, if the family relied on the sale of corn to buy store goods, it would need more than a single hectare (personal communication, 15 May 2002). In the 1980s Enge and Whiteford found that in Altepexi one hectare of irrigated maize produced 1,700 kg of grain, enough to feed a family of six with a little left over for the animals (1989, 47).

9   A young family of 5 consumes 5 maquilas (a measure of approximately 4 kg) of grain a week at 10 pesos each. This costs the family about 200 pesos a month (50 pesos × 4 weeks), or 2,600 pesos a year (50 pesos × 52 weeks). If we add the cost of feeding turkeys, pigs, and chickens, the annual cost in grain would be 2,600 plus 2,400 or 5,000 pesos in grain a year, equal to $545.50 based on the exchange rate of 1 peso = $0.1091 (February 2002).

10 Roberto González found in Oaxacan Sierra Juárez that male residents ate six to eight tortillas a day while women ate somewhat fewer (2001, 155). Tortillas must have weighed less or been smaller than those of rural Oaxaca in rural Morelos, where a recent study found that women eat five to fifteen tortillas daily and men fifteen to twenty or more. Factory workers ate considerably fewer tortillas (Martin, Cerullo, Bido, and Colmenares 2006).

11 In 2005 one maquila (approximately 4 kg) of Maseca yellow corn from elsewhere was sold in local stores for 7 pesos, or about $0.65. The same amount of local white criollo corn cost up to $1.10. In the following summer the same maquila of criollo corn cost 10 cents more. If residents want to save a little on their corn purchases, they buy a regional (but not local) variety called pitzahuac, a thin, hard corn which sells for slightly less than local criollo corn but tastes better and costs slightly more than Maseca corn.

12 Tarrant in Ajalpan, a major client of Levi's, responded to the formation of an independent union (SUITTAR) by firing two hundred union workers in 2003. Also, in late 2003 Martín Barrios Hernández, the leader of the Human and Labor Rights Commission of the Tehuacán Valley, was beaten and threatened ("Impune el hostigamiento a un defensor de trabajadores en Puebla," La Jornada, 9 February 2004). More recently Barrios Hernández was arrested on charges that he had asked a maquiladora owner, Lucio Gil Zárate, for cash in exchange for ending activism at his maquiladora. He was framed. Barrios's supporters have provided the state criminal judge with a video showing Barrios leading a demonstration in Ajalpan at the time when Gil Zárate claims that Barrios was at Gil's home in Tehuacán demanding money (La Jornada de Oriente, 5 January 2006; Maquiladora Solidarity Network, Urgent Action, 3 January 2006).

13 A newspaper article reported that only one-third of the seventeen industrial washers used by the maquiladoras have water treatment plants. The residual water contains chemical contaminants (*El Mundo de Tehuacán*, 9 May 2001). Barrios Hernández and Santiago Hernández report that there are over twenty-five industrial washers (2004, 87).

14 Interview, 30 May 2001, at the Tehuacán office of the Cámara nacional de la industria del vestido. This trend continues. In 2004 there were 150 registered maquiladoras in the city and two years later this number was down to 100.

15 Elisabeth Malkin, *New York Times*, 1 June 2009, cited in the Introduction.

16 An estimated 36.2 percent are now employed in United States industries, while only 13.3 percent work in agriculture (Delgado Wise 2006, 38).

17 Jeffrey Cohen (2001) summarizes the stages based on the work of Jones (1998) and D. S. Massey, L. Goldring, and J. Durand, "Continuities in Transnational Migration: An Analysis of Nineteen Mexican Communities," *American Journal of Sociology* 99, no. 6 (1994), 1492–1533.

18 For instance, the Coyoatl water society is a key political player in San José. Also, as noted earlier, throughout the twentieth century water socios often entered sharecropping and patron-client relationships with non-socios. Socios contribute monthly fees for the upkeep of the canals or for other water association expenses, but these represent only a fraction of the amount paid by those non-socios, who purchase or rent water from them or from associations in other towns. For example, in 2002 some water association members were paying 50 pesos monthly (per water share) for their electricity bills, plus other upkeep costs. To rent irrigation during the same year cost between 80 and 110 pesos an hour, totaling approximately 2,500 to 2,800 pesos for watering a milpa from cultivation to harvest.

19 All residents' names have been changed. Here I draw from five directed conversations with Ángela conducted on 2 June 2001, 17 November 2001, 2 June 2002, 30 May 2005, and 15 July 2006.

CHAPTER 5: From Campesinos to . . . ?

1 This is regionally varied. In their research in western Mexico, Zendajas and Mummert found that young migrants generally spoke about their futures in the United States, not in their hometowns. In practice this meant that if migrants inherited a parcel of land they would hire labor to work it in their absence. Most were uninterested in land purchase (1998, 181). It seems likely that those Sanjosepeños who stay in the United States for longer-than-average periods are more likely to envision and plan their future in their adopted country.

2 For the point about indigenous migrants Martínez Novo cites Alfonso Caso, *La comunidad indígena* (Mexico City: SEP-Diana, 1980).

3 Thanks to Ricardo F. Macip for helping me develop this idea. He makes a similar argument about indigenous producers asserting their political identity as campesinos in the neighboring Sierra Negra (Macip 1997, 2006).

4 From interviews at the SAGARPA office in Tehuacán, 9 May 2001. Interviews were conducted at the regional office and at the office in San Sebastián in 2002, 2003, and 2005.

5 All information on local maize production, varieties, and seed selection was gathered during interviews and field research. Regino Melchor Jiménez Escamilla helped to clarify my data on the different corn varieties.

6 Several residents mentioned this, including a store owner interviewed on 3 June 2005.

7 For a discussion of changing perspectives in agricultural research see Chambers, Pacey, and Thrupp eds. 1989. See also the summary by Glenn Stone (2007) of the innovation-diffusion theory and the comments section on his article, especially Mosse's comments on "rational scientists."

8 Michael Warren's definition (as discussed by Evelyn Mathias-Mundy with Gisele Morin-Labatut and Shahid Akhtar, "Background to the International Symposium on Indigenous Knowledge and Sustainable Development," *Indigenous Knowledge and Development Monitor* 1, no. 2 (1993), 2–5) cited in Gupta 1998, 173.

9 I owe this observation to a discussion with Leigh Binford.

10 This range is based on estimates from a former mayordomo, who said that costs were between 30,000 and 50,000 pesos for a year of service (July 2006).

11 Discussed in Georgina Saldierna's "Mientras más rebotan en EU a los migrantes indocumentados, más se afanan ellos en cruzar," *La Jornada*, 14 April 2001, 11. See also Cunningham 2004; De Genova 2005.

12 Thanks to Stephanie Bjork and Kathleen Bubinas for the phrase "the journey to work for women." They organized a panel by this title for the CASCA-SANA-UADY conference in Mérida, Mexico, in 2005, for which I presented a paper.

13 She cites Pierrette Hondagneu-Sotelo and Michael A. Messner, "The 'New Man' and Mexican Immigrant Man," *Theorizing Masculinities*, ed. Harry Brod and Michael Kaufman, 200–218 (Thousand Oaks: Sage, 1994), and Denise Kandiyoti, "Bargaining with Patriarchy," *Gender and Society* 2 (1988), 274–90.

14 Iztaccíhuatl is a volcano outside Mexico City, popularly called the Sleeping Lady, who is the heroine of myths and stories.

CONCLUSION

1 http://www.gruma.com/vEsp, accessed 6 May 2010.

2 "Llega hasta el zócalo la caravana Sin maíz no hay país, en contra del TLCAN," Jaime Avilés, *La Jornada*, 27 January 2008.

a medias sharecropping

acción a water share, typically five or six hours in the valley

agua sucia runoff, or "dirty" water

aguas broncas ravine water

atltepeame (Nahuatl) a twelve-hour share of the flow from
the Atzompa spring

atole a warm, traditional cornmeal-based beverage

avilacamachistas supporters of Governor Ávila Camacho in the 1930s

ayuntamiento municipal government or mayor's office

bajadas wells used to descend into underground canals (galerías)

bandoleros bandits

barranca ravine

barrio neighborhood

bienes comunales communal land

bracero temporary agricultural worker under the Bracero program
established by the United States (1942–64)

bulto bushel

cabecera head town or county capital

cacicazgo region ruled by a cacique

cacique indigenous leader or political leader

cajón a measure of grain equivalent to 4 kilograms; also a box

caliche hardpan or hardened deposit of calcium carbonate

campesino peasant

chichiltzi (Nahuatl) a criollo variety of red maize from the southern valley

chicuase (Nahuatl) a criollo variety of white maize with
a six-month growing cycle

cholo (modern usage) a person in the hip-hop scene or sporting hip-hop style; also, person of mixed racial heritage or Mexican-American; originally a derogatory term

científicos state officials who promoted modernization in the late nineteenth century

clase indígena indigenous people, or the indigenous social class

comal flat cooking pan or griddle

combis vans used for local transportation

compadrazgo system of ritual kinship

comunero communal landholder

co'tzi (Nahuatl) local variety of yellow maize

criollo(s) native and creolized varieties of maize

dotación a share of land or water granted under agrarian reform

encomienda colonial system whereby a tract of land and its inhabitants were placed under the charge of a colonialist

ejidatarios users of ejido land

ejido a land grant

ejote green bean

el campo the countryside

elote fresh corn or corn on the cob

faena communal labor

fiestas celebrations

galerías (filtrantes) underground chain wells

gente de razón non-Indians; literally, people of reason

gente del pueblo common folk; literally, people of the town or village

granjas farms; in the valley this refers to capital-intensive farms, agribusiness hatcheries, and poultry vaccine facilities

hacendados hacienda owners

haciendas colonial estate

huipil (Nahuatl) traditional blouse, often with embroidered pattern

indigenismo indigenist state policy and ideology

indígena indigenous

indigenista pro-indigenous government official or academic

indio(s) Indian

la clase blanca whites, or the white social class

leña firewood

lo mexicano that which is Mexican

macuiltzi (Nahuatl) a sweet criollo variety of maize

madrina godmother

maíz maize, corn

maíz transgénico transgenic or genetically engineered maize

manantiales natural springs

mandil apron traditionally worn by a peasant or indigenous woman

maquila short for maquiladora; also a regional measure for grain equal to approximately 4 kilograms

maquiladoras assembly plants or factories

maquilatitlán neologism meaning "place of many maquiladoras"

mariachis musicians and singers of popular Mexican songs

martesatl (Nahuatl) "Tuesday's water"; a share of the Atzompa spring

masa dough for making tortillas

mayordomía civil-religious cargo dedicated to the care of saints

mazorca corncob

mediador owner of land or water in a sharecropping relationship

mesa directiva neighborhood committee; also, board of directors

mestizaje racial intermixing

mestizo a person of mixed race or ethnicity

mezcal alcohol made from the agave plant

la migra United States border or immigration police

milpa cornfield

mojado/a "wetback" or undocumented migrant

molino mill

mozos agricultural day workers; also called peones

municipio county

muy guerreros very fierce or warrior-like

nahuitzi (Nahuatl) local rain-fed variety of corn with wide kernels and a four-month cycle

nixtamaleros people or mills which grind the nixtamal (corn soaked in lime and water) to make dough for tortillas

norteño, norteña northerner

originario, originaria an original inhabitant

padrino   godfather

pan de burro   a type of bread from San José; literally, "donkey bread"

pantles (Nahuatl)   a measure of land

peninsulares   Spaniards

peones   agricultural day workers; also called mozos

poblano, poblana   person from the state of Puebla

pocho   Mexican American; originally a derogatory term,
     sometimes used more neutrally

pozos   water wells

presidente municipal   mayor

pueblo   village, town, or people

quinceañeras   "sweet fifteen" party for teenage girls

ranchoatl (Nahuatl)   a twelve-hour share of the flow from
     the Atzompa spring

rastrear   to harrow or comb soil

reboso   a woven scarf traditionally worn by women

regidor(es)   councilperson(s)

repartimiento   colonial system of conscripted labor

Sanjosepeño, Sanjosepeña   person from San José

Semana Santa   Holy Week

sembrar   to plant

serrano   from the highlands

sexenio   six-year presidential term

sociedad   society or association

sociedades de la barranca   ravine water associations

sociedades explotadoras de agua   irrigation water associations

socios   shareholders or partners

surcado   to furrow the fields

taco   tortilla wrapped around a filling, usually meat

tamales   cornmeal-based filling wrapped in a corn husk or banana leaf

técnicos   agricultural extension workers

tecoatl (Nahuatl)   ancient canals; literally, "stone snakes"

tetlallí (Nahuatl)   traditional marriage broker and mediator

tetlallicihuatl (Nahuatl)   wife of marriage broker who also has
     mediation duties

tianguis (Nahuatl)   local or regional market

tortillería   tortilla makers, tortilla bakery

transgénico   transgenic; genetically engineered or modified

uso común   communal use

usuarios   users

yahuitzi (Nahuatl)   a variety of blue corn

yunta   yoke; steel plough pulled by oxen

zona de veda   restricted construction area

Bibliography

Aboites Aguilar, Luis. 1998. *El agua de la nación: Una historia política de México* (1888–1946). Mexico City: CIESAS.

Aboites Aguilar, Luis, et al., eds. 2000. *Fuentes para la historia de los usos del agua en México: 1710–1951*. Tlalpan, D.F.: Centro de Investigaciones y Estudios Superiores en Antropología Social.

Aguirre Beltrán, Gonzalo. 1967. *Regiones de Refugio*. Mexico City: INI.

———. 1992 [1986]. *Zongolica: Encuentro de dioses y santos patronos*. Mexico City: Fondo de Cultura Económica.

Aguirre Gómez, José Alfonso. 1999. "Análisis regional de la diversidad del maíz en el Sureste de Guanajuato." Doctoral thesis, Universidad Nacional Autónoma de México, Facultad de Ciencias, División de Estudios de Postgrado, Mexico City.

Aguirre Gómez, José Alfonso, Mauricio Bellon, and Melinda Smale. 1998. *A Regional Analysis of Maize Biological Diversity in Southeastern Guanajuato, Mexico*. Mexico City: CIMMYT.

Aitken, Rob. 1996. "Neoliberalism and Identity: Redefining State and Society in Mexico." *Dismantling the Mexican State?*, ed. R. Aitken, N. Craske, G. A. Jones, and D. Stansfield, 24–38. London: Macmillan.

Alonso, Ana María. 1995. *Thread of Blood*. Tucson: University of Arizona Press.

Alvarez-Morales, Ariel. 1999. "Mexico: Ensuring Environmental Safety While Benefiting from Biotechnology." *Agricultural Biotechnology and the Poor*, ed. Gabrielle J. Persley and Manuel M. Lantin, 90–96. Washington: Consultative Group on International Agricultural Research. http://www.cgiar.org/biotech/repo100/Morales.pdf.

Antal, Edit. 2007. Introduction. *Maize and Biosecurity in Mexico*, 1–10. Debate and Practice Issue of *Cuadernos del CEDLA*. Centre for Latin American Studies and Documentation, September.

Appadurai, Arjun. 1986. "Introduction: Commodities as the Politics of Value." *The Social Life of Things: Commodities in Cultural Perspective*, ed. A. Appadurai, 3–63. Cambridge: Cambridge University Press.

———. 1996. *Modernity at Large: Cultural Dimensions of Globalization*. Minneapolis: University of Minnesota Press.

Appendini, Kirsten. 1992. *De la milpa a los Tortibonos: La reestructuración de la política alimentaría en México*. Mexico City: El Colegio de México, UNRISD.

———. 1994. "Transforming Food Policy for Over a Decade: The Balance for Mexican Corn Farmers in 1993." *Economic Restructuring and Rural Subsistence in Mexico: Corn and the Crisis of the 1980s*, ed. C. Hewitt de Alcántara, 145–57. San Diego: United Nations Research Institute for Social Development (UNRISD)/Ejido Reform Research Project, Center for U.S.-Mexican Studies.

———. 1998. "Changing Agrarian Institutions: Interpreting the Contradictions." *The Transformation of Rural Mexico: Reforming the Ejido Sector*, ed. W. Cornelius and D. Myhre, 25–38. San Diego: Center for U.S.-Mexican Studies.

Aquino, Pedro. 1998. "Mexico." *Maize Seed Industries in Developing Countries*, ed. Michael Morris. Boulder: Lynne Rienner.

Araghi, Farshad. 2009. "The Invisible Hand and Visible Foot: Peasants, Dispossession and Globalization." *Peasants and Globalization*, ed. A. Haroon Akram-Lodhi and Cristóbal Kay, 111–47. New York: Routledge.

Arizpe, Lourdes. 1978. *Migración, etnicismo y cambio económico: Un estudio sobre migrantes campesinos a la ciudad de México*. Mexico City: Colegio de México.

———. 1981. "Relay Migration and the Survival of the Peasant Household." *Why People Move: Internal Migration and Development*, ed. J. Balan, 187–210. Paris: UNESCO.

Austin, James, and Gustavo Esteva. 1987. *Food Policy in Mexico. The Search for Self-Sufficiency*. Ithaca: Cornell University Press.

Bacon, David. 2008. "Displaced People: NAFTA's Most Important Product." *NACLA* 41, no. 5.

Baker, Lauren. 2007. "Regional Maize Marketing Initiatives." *The Maize and Biosecurity in Mexico*, 55–80. Debate and Practice Issue of *Cuadernos del CEDLA*. Centre for Latin American Studies and Documentation, September.

Barber, Pauline Gardiner, and Winnie Lem. 2004. "Commodities, Capitalism and Globalization." *Anthropologica* 46, no. 2, 123–29.

Barkin, David. 2002. "The Reconstruction of the Modern Mexican Peasantry." *Journal of Peasant Studies* 30, no. 1, 73–90.

———. 2003. "El maíz y la economía." *Sin Maíz, No Hay País*, ed. Gustavo Esteva and Catherine Marielle, 155–76. Mexico City: CONACULTA, Museo Nacional de Culturas Populares.

———. 2006. "Building a Future for Rural Mexico." *Latin American Perspectives* 33, no. 2, 132–40.

———, ed. 2006. *La gestión del agua urbana en México: Retos, debates y bienestar*. Guadalajara: Universidad de Guadalajara.

Barkin, David, and Billie De Walt, eds. 1990. *Food Crops vs. Feed Crops: Global Substitution of Grains in Production*. Boulder: Lynne Rienner.

Barkin, David, and Blanca Suárez. 1983. *El fin del principio: Las semillas y la seguridad alimentaria*. Mexico City: Centro de Ecología y Desarrollo.

Barrios Hernández, Martín, and Rodrigo Santiago Hernández. 2004. *Tehua-cán: Del calzón de manta a los blue jeans.* Tehuacán: Comisión de Derechos Humanos y Laborales del Valle de Tehuacán, in collaboration with the Maquila Solidarity Network, Canada.

Bartra, Armando. 1998. "Sobrevivientes: Historias en la frontera." *Globaliza-ción, crisis y desarrollo rural en América Latina,* 1–25. Memorías de V Congreso Latinamericano. Chapingo: Colegio de Posgraduados.

———. 2004. "Rebellious Cornfields: Toward Food and Labour Sovereignty." *Mexico in Transition: Neoliberal Globalism, the State, and Civil Society,* ed. Gerardo Otero, 18–36. London: Zed.

———. 2008. "The Right to Stay: Reactivate Agriculture, Retain the Popula-tion." *The Right to Stay Home: Alternatives to Mass Displacement and Forced Migration in North America,* 26–31. San Francisco: Global Exchange.

Bartra, Roger. 1974. *Estructura agraria y clases sociales en Mexico.* Mexico City: ERA-SEP. Engl. trans. Baltimore: Johns Hopkins University Press, 1993.

Basch, Linda, Cristina Blanc Szanton, and Nina Glick Schiller. 1995. *Nations Unbound: Transnational Projects, Postcolonial Predicaments, and Deterritorialized Nation-States.* New York: Gordon and Breach.

———. 1999. "From Immigrant to Transmigrant: Theorizing Transnational Migration." *Migration and Transnational Social Spaces,* ed. Ludger Pries, 72–105. Aldershot: Ashgate.

Beck, Ulrich. 1992 [1986]. *Risk Society: Towards a New Modernity.* London: Sage.

Bellon, Mauricio R., and Julien Berthaud. 2006. "Traditional Mexican Agri-cultural Systems and the Potential Impacts of Transgenic Varieties on Maize Diversity." *Agriculture and Human Values* 23, 3–14.

Bernstein, Henry. 2006. "Once Were/Still Are Peasants? Farming in a Glob-alising 'South.'" *New Political Economy* 11, no. 3, 399–406.

Binford, Leigh. 2002. "Social and Economic Contradictions of Rural Mi-grant Contract Labor between Tlaxcala, Mexico and Canada." *Culture and Agriculture* 24, no. 2, 1–19.

———. 2003. "Migrant Remittances and (Under) Development in Mexico." *Critique of Anthropology* 3, no. 3, 305–36.

———. 2004. "Lo local y lo global en la migración transnacional." *La economía política de la migración acelerada internacional de Puebla y Veracruz: Siete estudios de caso,* ed. Leigh Binford, 1–26. Mexico City: Luna Arena.

Bonfil Batalla, Guillermo. 1996 [1987]. *México Profundo: Reclaiming a Civiliza-tion.* Austin: University of Texas Press.

Boyer, Christopher. 2003. *Becoming Campesinos: Politics, Identity, and Agrarian Struggle in Postrevolutionary Michoacán, 1920–1935.* Stanford: Stanford University Press.

Brass, Tom. 1997. "The Agrarian Myth, the 'New' Populism and the 'New' Right." *Economic and Political Weekly* 32, no. 4, 27–42.

Brush, Stephen. 1996. "Whose Knowledge, Whose Genes, Whose Rights?"

Valuing Local Knowledge: Indigenous People and Their Intellectual Property Rights, ed. Stephen Brush and Doreen Stabinsky, 1–21. Washington: Island.

Brush, Stephen, and Doreen Stabinsky, eds. 1996. Valuing Local Knowledge: Indigenous People and Their Intellectual Property Rights. Washington: Island.

Buechler, Stephanie. 2005. "Women at the Helm of Irrigated Agriculture in Mexico: The Other Side of Male Migration." Opposing Currents, ed. Vivienne Bennett, Sonia Dávila-Poblete, and Maria Nieves Rico, 170–89. Pittsburgh: University of Pittsburgh Press.

Byers, Douglas. 1967. "The Region and Its People." The Prehistory of the Tehuacan Valley: Environment and Subsistence, ed. D. Byers, 34–47. Austin: University of Texas Press.

CECCAM, CENAMI, ETC Group, CASIFOP, UNOSJO, and AJAGI. 2003. "Contaminación transgénica del maíz en México: Mucho más grave." Press release, 9 October. http://www.etcgroup.org/article.asp?newsid=408.

Chambers, Robert, Arnold Pacey, and Lori Ann Thrupp, eds. 1989. Farmer First: Farmer Innovation and Agricultural Research. London: Intermediate Technology.

Chapela, Ignacio, and David Quist. 2001. "Trangenic DNA Introgressed into Traditional Maize Landraces in Oaxaca, Mexico." Nature 414, no. 6863, 541–43.

Chayanov, Alexander V. 1966. Theory of a Peasant Economy, ed. D. Thorner, B. Kerblay, and R. E. F. Smith. Homewood, Ill.: R.D. Irwin for the American Economic Association.

Cleveland, David, Daniela Soleri, and Flavio Aragon. 2003. "Transgenes on the Move." Paper presented at the American Anthropology Association Meetings in Chicago, 21 November 2003. Session 2-097.

Cleveland, David A., and Daniela Soleri. 2005. "Rethinking Risk Management Process for Genetically Engineered Crop Varieties in Small-Scale, Traditionally Based Agriculture." Ecology and Society 10, no. 1, 9. http://www.ecologyandsociety.org/vol10/iss1/art9.

Cleveland, David A., Daniela Soleri, Flavio Aragón Cuevas, José Crossa, and Paul Gepts. 2005. "Detecting (Trans)gene Flow to Landraces in Centers of Crop Origin: Lessons from the Case of Maize in Mexico." Environmental Biosafety Research 4, no. 4, 197–208.

Cohen, Jeffrey. 2001. "Transnational Migration in Rural Oaxaca, Mexico: Dependency, Development and the Household." American Anthropologist 103, no. 4, 954–67.

Cohen, Jeffrey, Richard Jones, and Dennis Conway. 2005. "Why Remittances Shouldn't Be Blamed for Rural Underdevelopment in Mexico." Critique of Anthropology 25, no. 1, 87–96.

Commission for Environmental Cooperation. 2004. Maize and Biodiversity: The Effects of Transgenic Maize in Mexico. Secretariat report. Montreal: North American Commission for Environmental Cooperation.

CONAPO (Consejo nacional de población). 2004. *Migración internacional.* http://www.conapo.gob.mx (accessed 5 January 2005).

Congreso Nacional Indígena. 2001. "Declaración por el reconocimiento constitucional de nuestros derechos colectivos." *Cuadernos Agrarios* 21 (Biopiratería y Bioprospección), 201–4.

Cook, Scott. 2006. "Commodity Cultures, Mesoamerica and Mexico's Changing Indigenous Economy." *Critique of Anthropology* 26, no. 2, 181–208.

Cook, Scott, and Leigh Binford. 1990. *Obliging Need: Rural Petty Industry in Mexican Capitalism.* Austin: University of Texas Press.

Cornelius, Wayne. 1998. "Ejido Reform: Stimulus or Alternative to Migration?" *The Transformation of Rural Mexico: Reforming the Ejido Sector,* ed. Wayne Cornelius and David Myhre, 229–46. San Diego: Center for U.S.-Mexican Studies.

Cornelius, Wayne, and David Myhre. 1998. Introduction. *The Transformation of Rural Mexico,* ed. Wayne Cornelius and David Myhre, 1–20. San Diego: Center for U.S.-Mexican Studies.

Corrigan, Philip, and Derek Sayer. 1985. *The Great Arch.* New York: Basil Blackwell.

Cotter, J. 1994a. "Before the Green Revolution: Agricultural Science Policy in Mexico, 1920–1950." PhD diss., Department of History, University of California, Santa Barbara.

———. 1994b. "The Origins of the Green Revolution in Mexico. Continuity or Change?" *Latin America in the 1940s: War and Postwar Transitions,* ed. D. Rock. Berkeley: University of California Press.

Cowen, M. P., and R. W. Shenton. 1996. *Doctrines of Development.* London: Routledge.

Crehan, Kate. 2002. *Gramsci, Culture and Anthropology.* Berkeley: University of California Press.

Cunningham, Hillary. 2004. "Nations Rebound? Crossing Borders in a Gated Globe." *Identities* 11, no. 3, 329–50.

Davis, Mike. 2000. *Magical Urbanism: Latinos Reinvent the US City.* London: Verso.

De Genova, Nicholas. 2005. *Working the Boundaries. Race, Space and "Illegality" in Mexican Chicago.* Durham: Duke University Press.

de Grammont, Hubert C. 1996. Introducción. *Neoliberalismo y organización social en el campo mexicano,* ed. H. C. de Grammont, 9–20. Mexico City: Plaza y Valdés.

de Janvry, Alain. 1981. *The Agrarian Question and Reformism in Latin America.* Baltimore: Johns Hopkins University Press.

de Janvry, Alain, Gustavo Gordillo de Anda, and Elisabeth Sadoulet. 1995. "NAFTA and Mexico's Maize Producers." *World Development* 23, no. 8, 1349–62.

———. 1997. *Mexico's Second Agrarian Reform: Household and Community Responses, 1990–1994.* San Diego: Center for U.S.-Mexican Studies.

de la Peña, Guillermo. 1988. "Los estudios regionales." *La antropología en México: Panorama histórico*, ed. M. Villalobos Salgado, 629–74. Mexico City: INAH.

Delborne, Jason Aaron. 2005. "Pathways of Scientific Dissent in Agricultural Biotechnology." PhD diss., Department of Environmental Science, Policy and Management, University of California, Berkeley.

Delgado Wise, Raúl. 2003. "Critical Dimensions of Mexico-U.S. Migration under the Aegis of Neoliberal Globalism." Paper presented at the conference International Migration in the Americas: Emerging Issues, 19–20 September 2003, York University. www.yorku.ca.cerlac.migration/Raul_Delgado.pdf (accessed 20 October 2003).

———. 2004. "Labour and Migration Policies under Vicente Fox: Subordination to U.S. Economic and Geopolitical Interests." *Mexico in Transition: Neoliberal Globalism, the State, and Civil Society*, ed. Gerardo Otero, 138–53. London: Zed.

———. 2006. "Migration and Imperialism: The Mexican Workforce in the Context of NAFTA" Trans. Mariana Ortega Breña, *Latin American Perspectives* 33, no. 2 (March), 33–45.

Delgado Wise, Raúl, and Humberto Márquez Covarrubias. 2008. "Capitalist Restructuring, Development and Labour Migration: The Mexico-US Case." *Third World Quarterly* 29, no. 7, 1359–74.

Desmarais, Annette-Aurélie. 2002. "The Vía Campesina: Consolidating an International Peasant and Farm Movement." *Journal of Peasant Studies* 29, no. 2, 91–124.

———. 2007. *La Vía Campesina: Globalization and the Power of Peasants*. Halifax: Fernwood.

de Teresa Ochoa, Ana Paula. 1996. "Una radiografía del minifundismo: población y trabajo en los valles centrales de Oaxaca (1930–1990)." *La sociedad rural mexicana frente al nuevo milenio*, ed. H. C. de Grammont and H. Tejera, 189–240. Mexico: UAM-INAH-UNAM Plaza y Valdes.

DeWalt, Billie, and Kathleen DeWalt. 1991. "The Results of Mexican Agriculture and Food Policy: Debt, Drugs, and Illegal Aliens." *Harvest of Want: Hunger and Food Security in Central America and Mexico*, ed. S. Whiteford and A. Ferguson, 189–207. Boulder: Westview.

Díaz León, Marco Antonio, and Artemio Cruz León, eds. 1998. *Nueve mil años de agricultura en México: Homenaje a Efraím Hernández Xolocotzi*. Mexico City: Grupo de Estudios Ambientales, A.C., Universidad Autónoma Chapingo.

Dirección General de Estadística. 1975. *Puebla: Censos agrícola-ganadero y ejidal 1970*. Mexico City: Dirección General de Estadística.

Douglas, Mary. 1966. *Purity and Danger: An Analysis of Concepts of Pollution and Taboo*. London: Routledge and Kegan Paul.

Dresser, Denise. 1991. *Neopopulist Solutions to Neoliberal Problems: Mexico's National Solidarity Program*. San Diego: Center for U.S.-Mexican Studies.

Durand, Jorge, Emilio Parrado, and Douglas Massey. 1996. "Migradollars and Development: A Consideration of the Mexican Case." *International Migration Review* 30, no. 2, 423–44.

Dyer, George, and Antonio Yúnez-Naude. 2003. "NAFTA and Conservation of Maize Diversity in Mexico." Report prepared for the Second North American Symposium on Assessing the Environmental Effects of Trade, Commission for Environmental Cooperation, 25–26 March 2003. http:// www.cec.org/files/PDF/ECONOMY/Dyer-Yunez_en.pdf.

Edelman, Marc. 1999. *Peasants against Globalization: Rural Social Movements in Costa Rica.* Stanford: Stanford University Press.

Embriz, Arnulfo, ed. 1993. *Indicadores socioeconómicos de los pueblos indígenas de México, 1990.* Mexico City: INI.

Enge, Kjell, and Scott Whiteford. 1989. *The Keepers of Water and Earth: Mexican Rural Social Organization and Irrigation.* Austin: University of Texas Press.

Escobar, Arturo. 1995. *Encountering Development.* Princeton: Princeton University Press.

——. 2001. "Culture Sits in Places: Reflections on Globalism and Subaltern Strategies of Localization." *Political Geography* 20, 139–74.

Esteva, Gustavo. 1983. *The Struggle for Rural Mexico.* South Hardy, Mass.: Bergin and Garvey.

——. 2003. "El maíz como opción de vida." *Sin maíz, no hay país,* ed. G. Esteva and C. Marielle, 285–322. Mexico City: CONACULTA/Museo de Culturas Populares.

Esteva, Gustavo, and Catherine Marielle, eds. 2003. *Sin maíz, no hay país.* Mexico City: CONACULTA/Museo de Culturas Populares.

ETC Group. 2007. "The World's Top 10 Seed Companies, 2006." http:www .etcgroup.org.

Ezcurra, Exequel, Sol Ortiz, and Jorge Soberón. 2002. "Evidence of Gene Flow from Transgenic Maize to Local Varieties in Mexico." *LMOs and the Environment: Proceedings of an International Conference, OECD.* Raleigh-Durham, the United States, 27–30 November 2001, 277–83.

Ferguson, James. 1994. *The Anti-Politics Machine: "Development," Depoliticization, and Bureaucratic Power in Lesotho.* Minneapolis: University of Minnesota Press.

Fitting, Elizabeth. 2004. " 'No hay dinero en la milpa': El maíz y el hogar transnacional del sur del Valle de Tehuacán." *La economía política de la migración acelerada internacional de Puebla y Veracruz: Siete estudios de caso,* ed. Leigh Binford, 61–101. Mexico City: Luna Arena.

——. 2006a. "Importing Corn, Exporting Labor: The Neoliberal Corn Regime, GMOs, and the Erosion of Mexican Biodiversity." *Agriculture and Human Values* 23, 15–26.

——. 2006b. "The Political Uses of Culture: Maize Production and the GM Corn Debates in Mexico." *Focaal: European Journal of Anthropology* 48, 17–34.

Fitzgerald, Ruth, and Hugh Campbell. 2001. "Food Scares and GM: Movement on the Nature/Culture Fault Line." *Australian Review of Public Affairs*. http://www.australianreview.net/digest/2001/10/fitzgerald_campbell .html (posted 5 October 2001, accessed 19 October 2002).

Flores Morales, Lourdes. 2008. *"No me gustaba, pero es trabajo": Mujer, trabajo y desechabilidad en la maquila*. Mexico City: Instituto de Ciencias Sociales y Humanidades "Alfonso Vélez Pliego," BUAP/Plaza y Valdés S.A. de C.V.

Fowler, Cary, and Pat Mooney. 1990. *Shattering: Food, Politics, and the Loss of Genetic Diversity*. Tucson: University of Arizona Press.

Fox, Jonathan. 1992. *The Politics of Food in Mexico: State Power and Social Mobilization*. Ithaca: Cornell University Press.

Fox, Jonathan, and Xóchitl Bada. 2008. "Migrant Organization and Hometown Impacts in Rural Mexico." *Journal of Agrarian Change* 8, nos. 2–3, 435–61.

Friedlander, Judith. 1975. *Being Indian in Hueyapan: A Study of Forced Identity in Contemporary Mexico*. New York: St. Martin's.

Friedmann, Harriet. 1987. "International Regimes of Food and Agriculture since 1870." *Peasants and Peasant Societies*, ed. Theodor Shanin, 258–76. Oxford: Basil Blackwell.

Fussell, Betty. 1992. *The Story of Corn*. New York: Alfred A. Knopf.

Gámez Andrade, Juan Manuel. 1998. "La revolución en nuestra tierra." *Tehuacán: Horizonte del Tiempo*, ed. E. Setién Gómez, 136–89. Tehuacán: Patrimonio Histórico de Tehuacán, A.C. Grupo Cagigas.

García-Barrios, Raúl, and Luis García-Barrios. 1990. "Environmental and Technological Degradation in Peasant Agriculture: A Consequence of Development in Mexico." *World Development* 18, no. 11, 1569–85.

———. 1994. "The Remnants of Community: Migration, Corn Supply and Social Transformation in the Mixteca Alta of Oaxaca." *Economic Transformation and Rural Subsistence in Mexico: Corn and the Crisis of the 1980s*, ed. C. Hewitt de Alcántara, 99–118. San Diego: Center for U.S.-Mexican Studies.

García Canclini, Néstor. 1990. *Culturas híbridas: estrategias para entrar y salir de la modernidad*. Mexico City: Grijalbo.

García García, Raymundo. 1998. *Puebla: Elecciones, legalidad y conflictos municipales, 1977–1995*. Puebla: BUAP y Dirección General de Fomento.

Gepts, Paul. 2002. "A Comparison between Crop Domestication, Classical Plant Breeding and Genetic Engineering." *Crop Science* 42, no. 6, 1780–90.

———. 2005. "Introduction of Transgenic Crops in Centers of Origin and Domestication." *Controversies in Science and Technology: From Maize to Menopause*, ed. A. Kinchy, D. Lee Kleinman, and J. Handelsman, 119–34. Madison: University of Wisconsin Press.

Gil Huerta, Gorgonio. 1972. "History of the Foundation of the Town of San Gabriel Chilacatla." *The Prehistory of The Tehuacan Valley: Chronology and Irrigation*, ed. F. Johnson, 154–61. Austin: University of Texas Press.

Gledhill, John. 1985. "The Peasantry in History: Some Notes on Latin American Research." *Critique of Anthropology* 5, no. 1, 33–56.

——. 1995. *Neoliberalism, Transnationalization and Rural Poverty: A Case Study of Michoacán, Mexico.* Boulder: Westview.

Glick Schiller, Nina. 2004. "Transnationality." *A Companion to the Anthropology of Politics*, ed. D. Nugent and J. Vincent, 448–67. Malden, Mass.: Blackwell.

Glowka, Lyle, Françoise Burhenne-Guilmin, and Hugh Synge. 1994. *A Guide to the Convention on Biological Diversity.* Gland, Switzerland: IUCN.

Goldberg, David Theo. 2002. *The Racial State.* Malden, Mass.: Blackwell.

Goldring, Luin. 1998. "Having Your Cake and Eating it Too: Selective Appropriation of Ejido Reform in Michoacán." *The Transformation of Rural Mexico*, ed. W. Cornelius and D. Myhre, 145–72. San Diego: Center for U.S.-Mexican Studies.

——. 1999. "Power and Status in Transnational Social Spaces." *Migration and Transnational Spaces*, ed. Ludger Pries, 162–86. Aldershot: Ashgate.

——. 2001. "The Gender and Geography of Citizenship in Mexico-U.S. Transnational Spaces." *Identities* 7, no. 4, 501–37.

González, Roberto. 2001. *Zapotec Science: Farming and Food in the Northern Sierra of Oaxaca.* Austin: University of Texas Press.

González Montes, Soledad. 1994. "Intergenerational and Gender Relations in the Transition from a Peasant Economy to a Diversified Economy." *Women of the Mexican Countryside, 1850–1990: Creating Spaces, Shaping Transitions*, ed. H. Folwer-Salamini and M. K. Vaughan, 175–91. Tucson: University of Arizona Press.

Goodman, David, and E. Melanie DuPuis. 2002. "Knowing Food and Growing Food: Beyond the Production-Consumption Debate in the Sociology of Agriculture." *Sociologia Ruralis* 42, no. 1, 5–22.

Greenberg, James. 1989. *Blood Ties: Life and Violence in Rural Mexico.* Tucson: University of Arizona Press.

Greenpeace Mexico. 2000. *Maíz transgénico: Documentos de campaña.* Mexico City: Greenpeace.

Grupo de Estudios Ambientales. 2004. *Foro en defensa del maíz nuestro: Por una agricultura campesina sustentable, sin transgénicos, en Guerrero.* Mexico City: UNORCA, UAG, GEA, Consejo Supremo de los Pueblos del Filo Mayor del Estado de Guerrero, CRESIG, CONACULTA.

——. 2007. *La contaminación transgénica del maíz en México: Luchas civiles en defensa del maíz y de la soberanía alimentaría.* Mexico City: Impretei.

Grupo de Estudios Ambientales et al. 2006. "Científicos y organizaciones no gubernamentales de México presentan el Manifiesto por la Protección de Maíz Mexicano." Press release, 25 June. www.gea-ac.org.

Gupta, Akhil. 1998. *Postcolonial Development: Agriculture and the Making of Modern India.* Durham: Duke University Press.

Gutiérrez, Natividad. 1999. *Nationalist Myths and Ethnic Identities: Indigenous Intellectuals and the Mexican State.* Lincoln: University of Nebraska Press.

Gutmann, Michael. 1996. *The Meanings of Macho: Being a Man in Mexico City.* Berkeley: University of California Press.

Hale, Charles. 2002. "Does Multiculturalism Menace? Governance, Cultural Rights, and the Politics of Identity in Guatemala." *Journal of Latin American Studies* 34, 485–524.

Hamilton, Sarah, Billie R. DeWalt, and David Barkin. 2003. "Household Welfare in Four Rural Mexican Communities: The Economic and Social Dynamics of Surviving National Crises." *Mexican Studies/Estudios Mexicanos* 19, no. 2, 422–51.

Hardin, Garrett. 1968. "The Tragedy of the Commons." *Science* 162, no. 3859, 1243–48.

Harvey, David. 1989. *The Condition of Postmodernity.* London: Basil Blackwell.

———. 2003. *The New Imperialism.* Oxford: Oxford University Press.

Hayden, Cori. 2003a. *When Nature Goes Public: The Making and Unmaking of Bioprospecting in Mexico.* Princeton: Princeton University Press.

———. 2003b. "From Market to Market: Bioprospecting's Idioms of Inclusion." *American Anthropologist* 30, no. 3, 359–71.

Heller, Chaia. 2002. "From Scientific Risk to Paysan Savoir-Faire: Peasant Expertise in the French and Global Debate over GM Crops." *Science as Culture* 11, no. 1, 5–37.

———. 2004. "Risky Science and Savoir Faire: Peasant Expertise in the French Debate over Genetically Modified Crops." *The Politics of Food,* ed. Marianne Elisabeth Lien and Brigitte Nerlich, 81–99. Oxford: Berg.

Heller, Chaia, and Arturo Escobar. 2003. "From Pure Genes to GMOs: Transnationalized Gene Landscapes in the Biodiversity and Transgenic Food Networks." *Genetic Nature/Culture: Anthropology and Science beyond the Two-Culture Divide,* ed. Alan H. Goodman, Deborah Heath, and M. Susan Lindee, 155–75. Berkeley: University of California Press.

Henao, Luis E. 1980. *Tehuacán: Campesinado e irrigación.* Mexico City: Edicol.

Henrich, Joseph. 2001. "Cultural Transmission and the Diffusion of Innovations: Adoption Dynamics Indicate That Based Cultural Transmission Is the Predominate Force in Behavioral Change." *American Anthropologist* 103, 992–1013.

Hernández Navarro, Luis. 2007. "The New Tortilla War." Americas Program Special Report, 7 May.

Herren, Ray V., and Roy L. Donahue. 1991. *The Agriculture Dictionary.* Albany, N.Y.: Delmar.

Herrick, Clare. 2005. " 'Cultures of GM': Discourses of Risk and Labeling of GMOs in the UK and EU." *Area* 37, no. 3, 286–94.

Hess, David. 1995. *Science and Technology in a Multicultural World.* New York: Columbia University Press.

Hewitt de Alcántara, Cynthia. 1976. *Modernizing Mexican Agriculture: Socioeconomic Implications of Technological Change, 1940–1970.* Geneva: UNRISD.

———. 1984. *Anthropological Perspectives on Rural Mexico.* London: Routledge.

———. 1994. "Economic Restructuring and Rural Subsistence in Mexico: Corn and the Crisis of the 1980s." *Ejido Reform Research Project.* San Diego: Center for U.S.-Mexican Studies/UNRISD.

Heyman, Josiah. 1994. "The Mexican-United States Border in Anthropology: A Critique and Reformulation." *Journal of Political Ecology* 1, 43–65.

Hindley, Jane. 1996. "Towards a Pluricultural Nation: The Limits of Indigenismo and Article 4." *Dismantling the Mexican State?*, ed. R. Aitken, N. Craske, G. A. Jones, and D. Stansfield, 225–43. London: Macmillan.

Hobsbawm, Eric. 1994. *Age of Extremes: The Short Twentieth Century, 1914–1991.* New York: Viking Penguin.

Holmes, Christina. 2006. "GMOs in the Laboratory: Objects without Everyday Controversy." *Focaal: European Journal of Anthropology* 48, 35–48.

Hoogvelt, Ankie. 2001. *Globalization and the Postcolonial World.* Baltimore: Johns Hopkins University Press.

INE-CONABIO. 2002. "Evidencias de flujo genético desde fuentes de maíz transgénico hacia variedades criollas." Paper presented by E. Huerta at the conference En Defensa del Maíz, Mexico City, 23 January 2002.

INEGI. 1990. *Anuario estadístico del estado de Puebla.*

———. 1994. *Puebla: Resultados definitivos VII censo ejidal.*

———. 1997. *Cuaderno estadístico municipal Tehuacán: Estado de Puebla, 1997.*

———. 2000. *Anuario estadístico, Puebla.*

———. 2005. *La migración de Puebla. XII Censo General de la Población y Vivienda 2000.* Mexico City.

ISAAA (International Service for the Acquisition of Agri-Biotech Applications). 2008. "Global Status of Commercialized Biotech/GM Crops, 2008: The First Thirteen Years, 1996–2008." Brief 39, executive summary. http://www. isaaa.org.

Jennings, Bruce. 1988. *Foundations of International Agricultural Research: Science and the Politics in Mexican Agriculture.* Boulder: Westview.

Jones, Richard C. 1992. "U.S. Migration: An Alternative Economic Mobility Ladder for Rural Central Mexico." *Social Science Quarterly* 73, no. 3, 496–510.

———. 1998. "Remittances and Inequality: A Question of Migration Stage and Geographic Scale." *Economic Geography* 74, no. 1, 8–25.

Kay, Cristóbal. 2008. "Reflections on Latin American Rural Studies in the Neoliberal Globalization Period: A New Rurality?" *Development and Change* 39, no. 6, 915–43.

Kearney, Michael. 1996. *Reconceptualizing the Peasantry.* Boulder: Westview.

Kenney, Martin. 1986. *Biotechnology: The University-Industrial Complex.* New Haven: Yale University Press.

Kloppenburg, Jack. 1988. *First the Seed: The Political Economy of Plant Biotechnology, 1492–2000.* Cambridge: Cambridge University Press.

Knight, Alan. 1990. "Racism, Revolution, and Indigenismo: Mexico, 1910–1940." *The Idea of Race in Latin America, 1870–1940*, ed. R. Graham, 71–113. Austin: University of Texas Press.

Kopytoff, Igor. 1986. "The Cultural Biography of Things: Commoditization as Process." *The Social Life of Things: Commodities in Cultural Perspective*, ed. Arjun Appadurai, 64–91. Cambridge: Cambridge University Press.

Kunz, Rachel. 2008. "Remittances Are Beautiful? Gender Implications of the New Global Remittance Trend." *Third World Quarterly* 29, no. 7, 1389–1409.

LaFrance, David G. 2003. *Revolution in Mexico's Heartland: Politics, War, and State Building in Puebla, 1913–1920.* Wilmington, Del.: Scholarly Resources.

Larson, Jorge, and Michelle Chauvet. 2004. "Understanding Complex Biology and Community Values: Communication and Participation." Background papers, Commission for Environmental Cooperation.

Latour, Bruno. 1987. *Science in Action: How to Follow Scientists and Engineers through Society.* Cambridge: Harvard University Press.

Lem, Winnie. 1988. "Household Production and Reproduction in Rural Languedoc: Social Relations of Petty Commodity Production in 'Broussan.'" *Journal of Peasant Studies* 15, no. 4, 500–529.

———. 1999. *Cultivating Dissent: Work, Identity, and Praxis in Rural Languedoc.* Albany, N.Y.: SUNY Press.

Lenin, Vladimir I. 1961 [1899]. "The Differentiation of the Peasantry." *The Development of Capitalism in Russia. Lenin's Collected Works*, vol. 3, 121–603. Moscow: Foreign Language Publishing House.

Levidow, Les. 1991. "Cleaning Up on the Farm." *Science as Culture* 2, part 4, no. 13, 538–68.

———. 2001. "Precautionary Uncertainty: Regulating GM Crops in Europe." *Social Studies of Science* 31, no. 6, 842–74.

Levidow, Les, and Joyce Tait. 1995. "The Greening of Biotechnology: GMOs as Environmentally Friendly Products." *Biopolitics: A Feminist and Ecological Reader on Biotechnology*, ed. V. Shiva and I. Moser, 121–38. London: Zed.

Lewontin, Stephen. 1983. "The Green Revolution and the Politics of Agricultural Development in Mexico since 1940." PhD diss., University of Chicago.

Lien, Marianne E. 2004. "The Politics of Food: An Introduction." *The Politics of Food*, ed. Marianne E. Lien and Brigitte Nerlich, 1–17. Oxford: Berg.

Lomelí Escalante, A. 1996. "El consumidor ante la controversia sobre la tortilla." *La industria de la masa y la tortilla*, ed. Felipe Torres et al. Mexico City: UNAM.

Lomnitz, Claudio. 2001. *Deep Mexico, Silent Mexico: An Anthropology of Nationalism.* Minneapolis: University of Minnesota Press.

Louette, Dominique. 1997. "Seed Exchange among Farmers and Gene Flow among Maize Varieties in Traditional Agricultural Systems." *Gene Flow among Maize Landraces, Improved Varieties and Teosinte*, ed. A. Serratos, M. Willcox, and F. Castillo-González. Mexico City: CIMMYT.

Macip, Ricardo F. 1997. "Politics of Identity and Internal Colonialism in the Sierra Negra of Mexico." MA thesis, Anthropology, New School for Social Research.

———. 2005. *Semos un país de peones: Café, crisis y el estado neoliberal en el centro de Veracruz.* Mexico City: ICSH, Benemérita Universidad Autónoma de Puebla.

MacNeish, Richard S. 1972. "Summary of the Cultural Sequence and Its Implications in the Tehuacan Valley." *The Prehistory of the Tehuacan Valley: Excavations and Reconnaissance*, ed. R. MacNeish, 496–504. Austin: University of Texas Press.

Mangelsdorf, Paul. 1974. *Corn: Its Origin, Evolution and Improvement*. Cambridge: Belknap.

Martin, Debora L., Margaret Cerullo, Jennifer N. Bido, and Jesus A. Colmenares. 2006. "Calcium Consumption and Bone Density in Factory and Farm Workers in Central Mexico." Paper presented at the meeting of the American Association of Physical Anthropologists, Anchorage, Alaska, 8–11 March.

Martínez Alier, Joan. 2001. "La lógica de la vida, no la del mercado." *Cuadernos Agrarios* 21 (Biopiratería y Bioprospección), 29–32.

Martínez Novo, Carmen. 2006. *Who Defines Indigenous? Identities, Development, Intellectuals, and the State in Northern Mexico*. New Brunswick: Rutgers University Press.

Marx, Karl. 1977. *Capital*, vol. 1. New York: First Vintage, Random House.

Massieu Trigo, Yolanda C. 1995. "La modernización biotecnológica de la agricultura mexicana: Otro sueño enterrado durante el sexenio salinista." *Cuadernos agrarios* 11–12 (December–January), 121–33.

Massieu Trigo, Yolanda, Rosa L. González, Michelle Chauvet, Yolanda Castañeda, and Rosa E. Barajas. 2000. "Transgenic Potatoes for Small-Scale Farmers: A Case Study in Mexico." *Biotechnology and Development Monitor* 41, March, 6–10.

Matsuoka, Yoshihiro, Yves Vigouroux, Major M. Goodman, Jesús G. Sánchez, Edward Buckler, and John Doebley. 2002. "A Single Domestication for Maize Shown by Multilocus Satellite Genotyping." *Proceedings from the National Academy of Science* 99, 6080–84.

McAfee, Kathleen. 2003a. "Corn Culture and Dangerous DNA: Real and Imagined Consequences of Maize Transgene Flow in Oaxaca." *Journal of Latin American Geography* 2, no. 1, 18–42.

———. 2003b. "Neoliberalism on the Molecular Scale: Economics and Genetic Reductionism in Biotechnology Battles." *Geoforum* 34, 203–19.

McGarity, Thomas. 2007. "Frankenfood Free: Consumer Sovereignty, Federal Regulation and Industry Control in Marketing and Choosing Food in the United States." *Labeling Genetically Modified Food*, ed. Paul Weirich, 128–50. Oxford: Oxford University Press.

McMichael, Philip. 2006. "Peasant Prospects in the Neoliberal Age." *New Political Economy* 11, no. 3, 407–18.

———. 2009. "A Food Regime Genealogy." *Journal of Peasant Studies* 36, no. 1, 139–69.

Mintz, Sidney. 1985. *Sweetness and Power: The Place of Sugar in Modern History*. New York: Viking.

Mitchell, Timothy. 2002. *Rule of Experts: Egypt, Techno-politics and Modernity*. Berkeley: University of California Press.

Mosse, David. 2005. *Cultivating Development: An Ethnography of Aid Policy and Practice*. London: Pluto.

Müller, Birgit. 2006a. "Infringing and Trespassing Plants: Patented Seeds at Dispute in Canada's Courts." *Focaal: European Journal of Anthropology* 48, 83–98.

——. 2006b. "On the Ownership of Nature." *Not for Sale: Decommodifying Public Life*, ed. Gordon Laxer and Dennis Soron, 55–68. Peterborough, Ont.: Broadview.

Mummert, Gail. 1994. "From *Metate* to *Despate*: Rural Mexican Women's Salaried Labor and the Redefinition of Gendered Spaces and Roles." *Women of the Mexican Countryside, 1850–1990*, ed. H. Fowler-Salamini and M. K. Vaughan, 192–209. Tucson: University of Arizona Press.

Murcott, Anne. 2001. "Public Beliefs about GM Foods: More on the Makings of a Considered Sociology." *Medical Anthropology Quarterly* 15, no. 1, 9–19.

Myhre, David. 1998. "The Achilles' Heel of the Reforms: The Rural Finance System." *The Transformation of Rural Mexico: Reforming the Ejido Sector*, ed. W. Cornelius and D. Myhre, 39–65. San Diego: Center for U.S.-Mexican Studies.

Nadal, Alejandro. 1999. "El maíz en México: Algunas implicaciones ambientales del Tratado de Libre Comercio de América del Norte." *Evaluación de los efectos ambientales del Tratado de Libre Comercio de América del Norte*, 66–182. Montreal: Commission for Environmental Cooperation.

——. 2000a. "Corn and NAFTA: An Unhappy Alliance." *Seedling: The Quarterly Newsletter of Genetic Resources Action International* 17, no. 2, 10–17.

——. 2000b. *The Environmental and Social Impacts of Economic Liberalization on Corn Production in Mexico*. Oxford: Oxfam Great Britain/World Wildlife Foundation.

Nader, Laura. 1996. Introduction. *Naked Science: Anthropological Inquiry into Boundaries, Power and Knowledge*, ed. L. Nader, 1–26. New York: Routledge.

National Research Council of the National Academies. 2002. *Environmental Effects of Transgenic Plants: The Scope and Adequacy of Regulation*. Washington: National Academies Press.

Newell, Peter. 2009. "Bio-Hegemony: The Political Economy of Agricultural Biotechnology in Argentina." *Journal of Latin American Studies* 41, 27–57.

Nugent, Daniel, and Ana María Alonso. 1994. "Multiple Selective Traditions in Agrarian Reform and Agrarian Struggle: Popular Culture and State Formation in the Ejido of Namiquipa, Chihuahua." *Everyday Forms of State Formation*, ed. G. Joseph and D. Nugent, 209–46. Durham: Duke University Press.

Nuijten, Monique. 2003. *Power, Community and the State: The Political Anthropology of Organisation in Mexico*. London: Pluto.

O'Brien, Jay, and William Roseberry. 1991. Introduction. *Golden Ages, Dark Ages: Imagining the Past in Anthropology and History*, ed. J. O'Brien and W. Roseberry, 1–18. Berkeley: University of California Press.

Ochoa, Enrique. 2000. *Feeding Mexico: The Political Uses of Food Since 1910.* Wilmington, Del.: Scholarly Resources.

Olivares Muñoz, F. 1995. *Estudio de mercado: Producción y comercialización del maíz elotero como hortaliza en la región de Tehuacán, Puebla.* MA thesis, Ingeniero Agrónomo, Economía Agrícola, Universidad Autónoma de Chapingo.

Ong, Aihwa. 1987. *Spirits of Resistance and Capitalist Discipline: Factory Women in Malaysia.* Albany, N.Y.: SUNY Press.

Ortega García, Sortero Jorge. 2001. *La vida y obra de Don Melitón Ramírez Valdez (1892–1937).* Tehuacán: GOMGAR.

Ortega-Paczka, Rafael. 1999. "Genetic Erosion in Mexico." *Proceedings of the Technological Meeting of the FAO World Information and Early Warning System on Plant Genetic Resources,* 69–75. Prague: Research Institute of Crop Production.

———. 2003. "La diversidad del maíz en México." *Sin maíz, no hay país,* ed. G. Esteva and C. Marielle, 123–43. Mexico City: CONACULTA, Museo Nacional de Culturas Populares.

Ortiz-García, Sol, E. Ezcurra, Bernd Schoel, Francisca Acevedo, Jorge Soberón, and Alison A. Snow. 2005. "Absence of Detectable Transgenes in Local Landraces of Maize in Oaxaca, Mexico (2003–2004)." *Proceedings of the National Academy of Sciences* 102, 12338–43.

Ortner, Sherry. 1984. "Theory in Anthropology Since the Sixties." *Comparative Studies in Society and History* 26, no. 1, 126–66.

Otero, Gerardo. 1999. *Farewell to the Peasantry? Political Class Formation in Rural Mexico.* Boulder: Westview.

Otero, Gerardo, Steffanie Scott, and Chris Gilbreth. 1997. "New Technologies, Neoliberalism, and Social Polarization in Mexico's Agriculture." *Cutting Edge: Technology, Information, Capitalism and Social Revolution,* ed. J. Davis, T. Hirschl, and M. Stack, 253–70. New York: Verso.

Pansters, Wil. 1990. *Politics and Power in Puebla: The Political History of a Mexican State, 1937–1987.* Amsterdam: CEDLA.

Paré, Luisa. 1977. *El proletariado agrícola en México: Campesinos sin tierra o proletarios agrícolas?* Mexico City: Siglo Veintiuno.

Paredes Colín, Joaquín. 1921. *El distrito de Tehuacán.* Tehuacán: El Refugio.

Paz, Octavio. 1961. *The Labyrinth of Solitude and Other Writings.* New York: Grove.

Pearson, Ruth, Ann Whitehead, and Kate Young. 1981. Introduction. *Of Marriage and the Market,* ed. K. Young, C. Wolkowitz, and R. McCullagh, ix–xix. London: Routledge.

Pechlaner, Gabriela, and Gerardo Otero. 2008. "The Third Food Regime: Neoliberal Globalism and Agricultural Biotechnology in North America." *Sociologia Ruralis* 48, no. 4, 1–21.

Perales, Hugo, Bruce F. Benz, and Stephen B. Brush. 2005. "Maize Diversity and Ethnolinguistic Diversity in Chiapas, Mexico." *Proceedings of the National Academy of Sciences* 102, no. 3, 949–54.

Pérez, Mamerto, Sergio Schlesinger, and Timothy Wise. 2008. *The Promise* and Perils of Agricultural Trade Liberalization. Lessons from Latin America. Washington: Washington Office on Latin America.

Pilcher, Jeffrey. 1998. ¡Que vivan los tamales! Food and the Making of Mexican Identity. Albuquerque: University of New Mexico Press.

Plan Puebla. 1974. "The Puebla Project: Seven Years of Experience, 1967–1973: Analysis of a Program to Assist Small Subsistence Farmers to Increase Crop Production in a Rainfed Area of Mexico." El Batan, Mexico: CIMMYT.

Pollan, Michael. 2006. The Omnivore's Dilemma: A Natural History of Four Meals. New York: Penguin.

Poole, Deborah. 1994. "Anthropological Perspectives on Violence and Culture: A View from the Peruvian High Provinces." Unruly Order, ed. D. Poole, 1–29. Boulder: Westview.

———. 1997. Vision, Race and Modernity: A Visual Economy of the Andean World. Princeton: Princeton University Press.

Prakash, Channapatna S. 2005. "Duh . . . No GM Genes in Mexican Corn." AgBioWorld. http://www.agbioworld.org/newsletter_wm/index.php?case id=archive&newsid=2398 (posted 9 August 2005, accessed 10 October 2005).

Pred, Allan, and Michael Watts. 1992. Reworking Modernity: Capitalisms and Symbolic Discontent. New Brunswick: Rutgers University Press.

Pries, Ludger. 1999. "New Migration in Transnational Spaces." Migration and Transnational Social Spaces, ed. L. Pries. Aldershot: Ashgate.

Quist, David, and Ignacio H. Chapela. 2002. "Biodiversity (Communications Arising (Reply)): Suspect Evidence of Transgenic Contamination/Maize Transgene Results in Mexico Are Artefacts." Nature 416, no. 6881, 602.

Razack, Sherene. 2002. Race, Space and the Law: Unmapping a White Settler Colony. Toronto: Between the Lines.

Redclift, Michael. 1980. "Agrarian Populism in Mexico: The 'Via Campesina.'" Journal of Peasant Studies 7, no. 4, 492–502.

Ribeiro, Silvia. 2003. "México, caballo de Troya de los transgénicos en América Latina." La Jornada, 27 December.

———. 2004. "The Day the Sun Dies: Contamination and Resistance in Mexico." Seedling, July, 5–10.

Richards, Paul. 1989. "Agriculture as Performance." Farmer First: Farmer Innovation and Agricultural Research, ed. R. Chambers, A. Pacey, and L. A. Thrupp, 39–51. London: Intermediate Technology.

Rimarachín, Isidro Cabrera, Emma Zapata Martelo, and Verónica Vázquez García. 2001. "Gender, Rural Households, and Biodiversity in Native Mexico." Agriculture and Human Values 18, 85–93.

Rissler, Jane, and Margaret Mellon. 1996. The Ecological Risks of Engineered Crops. Cambridge: MIT Press.

Rivermar Pérez, María Leticia. 2000. "La reconstrucción de las identidades sociales en el contexto de las migraciones." Conflictos migratorios trans-

nacionales y respuestas comunitarias, ed. L. Binford and M. E. D'Aubeterre, 81–96. Mexico City: BUAP.

Roseberry, William. 1989. *Anthropologies and Histories*. New Brunswick: Rutgers University Press.

——. 1993. "The Agrarian Question." *Confronting Historical Paradigms: Peasants, Labor, and the Capitalist World System in Africa and Latin America*, ed. F. Cooper et al. Madison: University of Wisconsin Press.

——. 1994. "Hegemony and the Language of Contention." *Everyday Forms of State Formation*, ed. G. Joseph and D. Nugent, 355–66. Durham: Duke University Press.

——. 1996. "Hegemony, Power, and Languages of Contention." *The Politics of Difference*, ed. E. Wilmsen and P. McAllister, 71–84. Chicago: University of Chicago Press.

Rouse, Roger. 1991. "Mexican Migration and the Social Space of Postmodernism." *Diaspora* 1, no. 1, 8–23.

Rubin, Jeffrey. 1997. *Decentering the Regime: Ethnicity, Radicalism, and Democracy in Juchitán, Mexico*. Durham: Duke University Press.

SAGARPA–SIAP. 2004. *Situación actual y perspectivas del maíz en México, 1990–2004*. Mexico City: SAGARPA-SIAP.

——. 2006. *Valorización de la cadena agroalimentaria maíz grano: Diagrama de flujo*. Web site, accessed 10 February 2008.

——. 2007. *Servicio de alimentación agroalimentaria y pesquera*. www.siap.gob.mx (accessed 10 February 2008).

Sain, Gustavo, and Miguel López-Pereira. 1999. *Maize Production and Agricultural Policies in Central America and Mexico*. Mexico City: CIMMYT.

Salazar Exaire, Celia. 2000. *Uso y distribución del agua en el valle de Tehuacán. El caso de San Juan Bautista Axalpan, Pue (1610–1798)*. Serie Antropología Social. Mexico City: INAH.

Schryer, Frans. 1990. *Ethnicity and Class Conflict in Rural Mexico*. Princeton: Princeton University Press.

Schurman, Rachel. 2003. "Introduction: Biotechnology and the New Millennium." *Engineering Trouble: Biotechnology and its Discontents*, ed. Rachel Schurman and Dennis Doyle Takahashi Kelso, 1–23. Berkeley: University of California Press.

Scott, James. 1976. *The Moral Economy of the Peasant: Rebellion and Subsistence in Southeast Asia*. New Haven: Yale University Press.

——. 1985. *Weapons of the Weak: Everyday Forms of Peasant Resistance*. New Haven: Yale University Press.

——. 1998. *Seeing like a State: How Certain Schemes to Improve the Human Condition Have Failed*. New Haven: Yale University Press.

SENASICA. 2005. "*Ensayos de productos genéticamente modificados autorizados en México de 1988 al 11 de octubre de 2005.*" SAGARPA.

Serratos, José Antonio. 1996. "Evaluation of Novel Crop Varieties in Their Center of Origin and Diversity: The Case of Maize in Mexico." Turning

Priorities into Feasible Programs. Proceedings of a Policy Seminar on Agricultural Biotechnology for Latin America, Lima, Peru, 6–10 October 1996. Agricultural Biotechnology Policy Seminars no. 4, 68–73.

Serratos, José Antonio, José-Luis Gómez-Olivares, Noé Salinas-Arreortua, Enrique Buendía-Rodríguez, Fabián Islas-Gutiérrez, and Ana de Ita. 2007. "Transgenic Proteins in Maize in the Soil Conservation Area of Federal District, Mexico." *Frontiers in Ecology* 5, no. 5, 247–52.

Serratos, José Antonio, Martha Willcox, and Fernando Castillo-González, eds. 1996. *Flujo genético entre maíz criollo, maíz mejorado y teocintle: implicaciones para el maíz transgénico* Mexico City: CIMMYT [Engl. trans. published 1997].

Shiva, Vandana. 1991. *The Violence of the Green Revolution.* London: Zed.

———. 1993. *Monocultures of the Mind: Perspectives on Biodiversity and Biotechnology.* London: Zed.

Smith, Gavin A. 1991. "The Production of Culture in Local Rebellion." *Golden Ages, Dark Ages: Imagining the Past in Anthropology and History,* ed. W. Roseberry and J. O'Brien, 180–207. Berkeley: University of California Press.

Smith, Richard C. 1998. "Transnational Localities: Community, Technology and the Politics of Membership within the Context of Mexico and U.S. Migration." *Transnationalism from Below: Comparative Urban and Community Research* 6, ed. Michael Peter Smith and Luis Eduardo Guarnizo, 196–239.

Snow, Allison. 2005. "Genetic Modification and Gene Flow: An Overview." *Controversies in Science and Technology: From Maize to Menopause,* ed. A. Kinchy, D. Lee Kleinman, and J. Handelsman, 107–18. Madison: University of Wisconsin Press.

Soleri, Daniela, and David A. Cleveland. 2006. "Transgenic Maize and Mexican Maize Diversity: Risky Synergy?" *Agriculture and Human Values* 23, 27–31.

Soleri, Daniela, David A. Cleveland, Flavio Aragón, Mario R. Fuentes, Humberto Ríos, and Stuart H. Sweeney. 2005. "Understanding the Potential Impact of Transgenic Crops in Traditional Agriculture: Maize Farmers' Perspectives in Cuba, Guatemala, and Mexico." *Environmental Biosafety Research* 4, no. 3, 141–66.

Soleri, Daniela, David Cleveland, and Flavio Aragón Cuevas. 2006. "Transgenic Crops and Crop Varietal Diversity: The Case of Maize in Mexico." *Bioscience* 56, no. 6, 503–13.

Solís, Felipe. 1998. *La cultura de maíz.* Mexico City: Clío, Libros y Videos.

Stavenhagen, Rodolfo. 1969. *Las clases sociales en las sociedades agrarias.* Mexico City: Siglo XXI.

Stédile, João Pedro. 2002. "Landless Battalions" (interview). *New Left Review* 15, 77–104.

Stephen, Lynn. 1998. "Interpreting Agrarian Reform in Two Oaxacan Ejidos: Differentiation, History, and Identities." *The Transformation of Rural*

Mexico: Reforming the Ejido Sector, ed. W. Cornelius and D. Myhre, 125–44. San Diego: Center for U.S.-Mexican Studies.

———. 2005. Zapotec Women: Gender, Class and Ethnicity in Globalized Oaxaca. 2nd edn. Durham: Duke University Press.

Stone, Glenn Davis. 2002. "Both Sides Now. Fallacies in the Genetic-Modification Wars, Implications for Developing Countries, and Anthropological Perspectives." Current Anthropology 43, no. 4, 611–19.

———. 2007. "Agricultural Deskilling and the Spread of Genetically Modified Cotton in Warangal." Current Anthropology 48, no. 1, 67–87.

Stone, M. Priscilla, Angelique Haugerud, and Peter D. Little. 2000. "Commodities and Globalization: Anthropological Perspectives." Commodities and Globalization, ed. A. Haugerud, M. P. Stone, and P. Little. New York: Rowman and Littlefield.

Takacs, David. 1996. The Idea of Biodiversity: Philosophies of Paradise. Baltimore: Johns Hopkins University Press.

Thies, Janice E., and Medha H. Devare. 2007. "An Ecological Assessment of Transgenic Crops." Transgenics and the Poor: Biotechnology in Developing Countries, ed. Ronald Herring, 97–129. New York: Routledge.

Tsing, Anna. 2000. "The Global Situation." Cultural Anthropology 15, no. 3, 327–60.

Turrent, Antonio, Rodrigo Aveldaño Salazar, and Rodolfo Moreno Dahme. 1996. "Análisis de las posibilidades técnicas de la autosuficiencia sostenible de maíz en México." TERRA 14, no. 4, 445–68.

Turrent, Antonio, and J. Antonio Serratos. 2004. "Context and Background on Maize and Its Wild Relatives in Mexico." Maize and Biodiversity: The Effects of Transgenic Maize in Mexico. Commissioned for the Secretariat of the Commission on Environmental Cooperation. http://www.cec.org/maize/resources/chapters.cfm?varlan=english.

Turrent Fernández, Antonio. 2005. "La diversidad genética del maíz y del teocintle de México debe ser protegida contra la contiminación irreversible del maíz transgénico." Transgénicos, quien los necesita?, ed. Armando Bartra et al., 51–60. Mexico City: Grupo Parlamentario del PRD den la LIX Legislatura, Centro de Producción Editorial.

U.S. Department of Agriculture, Economic Research Service. 2009. "Adoption of Genetically Engineered Crops in the U.S.: Corn Varieties (2000–2009)." http://www.ers.usda.gov/data/biotechcrops/ExtentofAdoption Table1.htm.

Vasconcelos, José. 1997 [1925]. La raza cósmica. Trans. and with an introd. by Didier T. Jaén. Baltimore: Johns Hopkins University Press.

Vaughan, Mary Kay. 1997. Cultural Politics in Revolution: Teachers, Peasants, and Schools in Mexico, 1930–1940. Tucson: University of Arizona Press.

Vera Herrera, Ramón. 2004. "In Defense of Maize (and the Future)." Citizen Action in the Americas 13 (August), 1–10. Program of the Americas, Interhemispheric Resource Center (IRC).

Vizcarra Bordi, Ivonne. 2006. "The 'Authentic' Taco and Peasant Woman: Nostalgic Consumption in the Era of Globalization." *Culture and Agriculture* 28, no. 2, 97–107.

Walsh, Casey, and Elizabeth Ferry. 2003. "Introduction: Production, Power and Place." *The Social Relations of Mexican Commodities: Power, Production and Place*, ed. Casey Walsh et al., 1–18. San Diego: Center for U.S.-Mexican Studies.

Warman, Arturo. 1980 [1976]. *We Come to Object: The Peasants of Morelos and the National State*. Baltimore: Johns Hopkins University Press.

——. 2003 [1988]. *Corn and Capitalism: How a Botanical Bastard Grew to Global Dominance*, trans. Nancy Westrate. Chapel Hill: University of North Carolina Press.

Whatmore, Sarah, and Lorraine Thorne. 1997. "Nourishing Networks: Alternative Geographies of Food." *Globalising Food: Agrarian Questions and Global Restructuring*, ed. D. Goodman and M. Watts, 287–304. London: Routledge.

Whiteford, Scott, and Francisco Bernal. 1996. "Campesinos, Water and the State: Different Views of La Transferencia." *Reforming Mexico's Agrarian Reform*, ed. L. Randell, 223–34. New York: Sharpe.

Whiteford, Scott, Francisco A. Bernal, Heliodoro Díaz Cisneros, and Esteban Valtierra-Pacheco. 1998. "Arid-Land Ejidos: Bound by the Past, Marginalized by the Future." *The Transformation of Rural Mexico: Reforming the Ejido Sector*, ed. W. Cornelius and D. Myhre, 381–99. San Diego: Center for U.S.-Mexican Studies.

Whiteford, Scott, and Roberto Melville, eds. 2002. *Managing a Sacred Gift: Changing Water Management Strategies in Mexico*. San Diego: Center for U.S.-Mexican Studies.

Wilkes, Garrison. 2007. "Urgent Notice to All Maize Researchers: Disappearance and Extinction of the Last Wild Teosinte Population Is More Than Half Completed." *Maydica* 52.

Williams, Raymond. 1961. *The Long Revolution*. New York: Harper and Row.

——. 1973. *The Country and the City*. Oxford: Oxford University Press.

Wise, Timothy. 2007. "Policy Space for Mexican Maize: Promoting Agrobiodiversity by Promoting Rural Livelihoods." Global Development and Environment Institute Working Paper no. 07-01.

Wolf, Eric. 1955. "Types of Latin American Peasantry: A Preliminary Discussion." *American Anthropologist* 57, no. 3, 452–71.

——. 1982. *Europe and the People without History*. Berkeley: University of California Press.

——. 1986. "The Vicissitudes of the Closed Corporate Peasant Community." *American Ethnologist* 13, no. 2, 325–29.

Woodbury, Richard, and James Neely. 1972. "Water Control Systems of the Tehuacan Valley." *The Prehistory of the Tehuacan Valley*, ed. F. Johnson, 81–153. Austin: University of Texas Press.

Worthy, Kenneth A., Richard C. Strohman, Paul R. Billings, Jason A. Delborne, Earth Duarte-Trattner, Nathan Gove, Daniel R. Latham, and Carol M. Manahan. 2005. "Agricultural Biotechnology Science Compromised: The Case of Quist and Chapela." *Controversies in Science and Technology: From Maize to Menopause*, ed. Daniel Lee Kleinman, Abby J. Kinchy, and Jo Handelsman. Madison: University of Wisconsin Press.

Wynne, Brian. 2001. "Creating Public Alienation: Expert Cultures of Risk and Ethics on GMOs." *Science as Culture* 10, no. 4, 445–81.

Zahniser, Steven, and William Coyle. 2004. *U.S.-Mexico Corn Trade during the NAFTA Era: New Twists to an Old Story*. Electronic Outlook Report from the Electronic Research Service. FDS-04D-01, May 2004. www.ers.usda.gov.

Zendejas, Sergio, and Gail Mummert. 1998. "Beyond the Agrarian Question: The Cultural Politics of Ejido Natural Resources." *The Transformation of Rural Mexico: Reforming the Ejido Sector*, ed. W. Cornelius and D. Myhre, 173–202. San Diego: Center for U.S.-Mexican Studies.

Maize (cont.)
60, 253 n. 18; subsidies for, 17, 65, 230; as symbol of Mexico, 1, 14, 33, 36, 65, 67, 74, 76–77, 79, 83, 116, 232; valley narratives about, 6, 188–95; yields of, 12, 18, 36, 41, 53–56, 58, 72, 75, 89, 90, 92, 94, 103–4, 108, 119, 169, 205, 253 n. 19, 256 n. 2, 260 n. 2, 260 n. 5. See also Criollos; Elote; Food; Milpas; Tortillas

Maize varieties, 15–16; center of biodiversity of, 43; classification of, 90; decline in, 12, 44, 159, 194, 204–5; genetically engineered or modified, 47–48, 55; "modern" vs. "traditional," 40–42, 44, 53, 108, 207, 253 n. 19; in San José, 152, 159, 194, 203–5, 209–10, 261 n. 11. See also Criollos; Seed

Maquiladoras, 8, 12, 32, 109, 176, 179–80, 185, 187, 213–14, 217, 230, 261 n. 12, 262 n. 14; environmental problems due to, 124, 178; gendered work and, 21, 31, 156–57, 177, 189–90, 195, 200, 211, 215, 219, 220–24; mobility of workers and, 198–200; unions and, 178; water use and, 124, 262 n. 13; workers in, 6, 11, 13–14, 25, 31, 126, 156–58, 175, 186, 226–28, 232; workers as disposable labor in, 177, 195, 229, 231; working conditions in, 158, 177–78

Maquilas, 261 n. 9, 261 n. 11

Markets, 8, 13–14, 18, 20, 26, 30, 54, 76, 80, 87, 96, 99, 103, 107–8, 115; efficiency of, 4, 36, 74, 232; free, 21, 29, 102, 112, 160, 162; intermediaries and, 83, 105–6; for labor, 24, 100, 157, 175, 234, 236; for seed, 41–44, 204–5, 208; supermarkets, 17, 19; in the valley, 11, 16, 25, 133, 151, 159, 169, 173, 175–76, 186–87, 192, 194, 203–4, 211, 219, 223; for water, 134; world, 22, 98, 104, 114

Marriage, 121, 186, 189, 211, 219–20, 222–23; marriage brokers, 126, 132, 139, 149, 258 n. 12

Martínez Novo, Carmen, 199, 225

Masculinity, 216–18. See also Gender

Mayordomía (civil-religious cargo), 97, 120–21, 123–24, 126–27, 129, 131–34, 142, 180, 216

Mestizaje, 78–79, 82, 87–89

Mestizo, 67, 79–81, 84, 87, 90, 94, 121, 123, 127–28, 138, 147, 176, 199, 221, 227–28; as "Cosmic Race," 82; defined, 77; as symbol of modern nation, 14, 78, 82–83, 88–89, 116

Mexican revolution, 23, 31, 81–82, 85, 88, 138–39, 202

Migrants: agriculture disfavored by, 181, 203, 208–9, 228, 231, 236; consumption patterns among, 183–87, 214–16, 227; as deportable labor, 182, 224; as disposable labor, 12–13, 175, 182, 229, 231; ethnicity and, 224–27; gender and, 84, 157, 179–81, 200, 212–14, 216–18, 221–22; generational differences among, 12, 25, 126, 156–59, 175, 180–81, 190, 193, 197, 199–200, 209–10, 217, 220, 227–28; mobility of, 217–19, 231; San José and, 118, 180–81, 185–95, 198, 214, 224, 232; social costs of migration and, 182, 211, 213, 227, 231; social prestige of, 183, 186–87, 200, 216–17, 226–28; theories of migration and, 21, 23–24, 100, 182–85, 230–31; transnational, 118, 157, 180–81, 186–88, 200, 230–31, 235. See also Braceros; Coyotes; Remittances

Milpas: gendered labor of, 12, 84, 117, 120, 133, 135, 165–66, 188, 211; in Tehuacán Valley, 8, 12, 114, 120, 136, 155–56, 158, 173, 175, 190, 193–94, 198, 209–10, 228; traditional, 8, 12, 114, 159, 167; transnational, 180–81, 186–87

Ministry of Agriculture. *See* SAGARPA

Monsanto, 42, 45, 50, 63, 71, 72, 110, 251 n. 5, 252 n. 13, 253 n. 15, 255 n. 28, 255 n. 30

Nadal, Alejandro, 71, 254 n. 22

NAFTA (North American Free Trade Agreement), 11, 17, 19–20, 29, 37–38, 46–49, 65–66, 69, 101–3, 109, 115, 158, 160, 176, 230, 233–34, 256 n. 4. *See also* CEC; Trade liberalization

Nahuatl language, 8, 121, 125, 131, 177, 199, 204, 209, 225–26

Neoliberal corn regime, 4–5, 19, 38, 41–42, 56, 73, 75, 92–93, 101–8, 116, 158, 163, 194–95 200, 229, 232

Neoliberalism, 18, 26, 28, 31, 33, 76–77, 137; as approach to development, 20, 103–4, 177, 184, 195, 200; governance and, 48–49, 233; irrigation management and, 146, 259 n. 22; opposition to, 24, 29, 66, 109–12, 116, 234; peasants transformed by, 4, 14, 24, 199, 202; structural adjustment and, 4, 6; in the valley, 5, 11, 118, 156, 233

"Neoliberal multiculturalism," 233

NGOs (nongovernmental organizations and civil associations), 56, 59–61, 65–66, 68–70, 72–73, 110–11, 113, 255 n. 31

Nixtamal 16, 84, 93, 103

Oaxaca, 5, 7, 20, 70, 112, 118, 146, 149–52, 179, 217, 219, 261 n. 10; GM maize and transgenes in, 35, 37, 50, 53, 56, 59–60, 65, 67, 69, 121, 233, 253 n. 20, 254 nn. 25–26

Organic agriculture, 110, 113, 249 n. 2

Otero, Gerardo, 19

Peasantry: dispossession and, 30–31; peasant essentialism, 13–14, 22, 76, 114–16, 233; rights movements of, 6, 63, 76, 111. *See also* Campesinos

Pechlaner, Gabriela, 19

Pilcher, Jeffrey, 80, 94

Place, 15, 28, 31, 118, 236

Plan Puebla Panama, 177

Political economy, 1, 27, 31, 111, 250 n. 11

Political movements, 2; agrarian, 138, 142; alter-globalization, 63, 66, 97, 111, 114–15; anti-GM, 57, 65–66, 75; indigenous, 89, 118, 224, 233, 235; peasant, 6, 63, 76, 111; transnational, 76, 116; transnational food sovereignty, 5, 110–12

Political parties: PAN (National Action Party), 20, 144–45, 250 n. 12; PRD (Party of the Democratic Revolution), 10, 145, 249 n. 6, 259 n. 19; PRI (Institutional Revolutionary Party), 10, 20, 86, 144–45, 154, 162, 249 nn. 5–6, 256 n. 4; PSUM (Unified Socialist Party of Mexico), 144–45, 154, 249 n. 6, 259 n. 19

Politics, 4, 20, 38; in San José, 137–38, 143–45, 259 n. 19, 262 n. 18. *See also* State

Pollan, Michael, 46

Porfiriato, 79–82, 85–86, 88, 98, 128

Porfirio Díaz, José de la Cruz, 79, 128

Precautionary principle, 48–49, 55, 71–72, 252 n. 9

Pred, Allan, 26, 28, 118

Primitive accumulation, 26, 30. *See also* Accumulation by dispossession

Procampo, 17, 103, 109, 160–62, 169, 187, 193, 203

Procede (Program for the Certification of Ejido Rights and the Titling of Urban House Plots), 137, 160–61

ELIZABETH FITTING
is an associate professor in the Department of Sociology
and Social Anthropology at Dalhousie University.

Library of Congress Cataloging-in-Publication Data

Fitting, Elizabeth M.
The struggle for maize : campesinos, workers, and transgenic
corn in the Mexican countryside / Elizabeth Fitting.
p. cm.
Includes bibliographical references and index.
ISBN 978-0-8223-4928-9 (cloth : alk. paper)
ISBN 978-0-8223-4956-3 (pbk. : alk. paper)
1. Corn industry—Mexico.
2. Corn—Genetic engineering—Mexico.
3. Transgenic plants—Mexico. I. Title.
HD9049.C8M4796 2011
338.1'75670972—dc22 2010028812